T0233923

Nanotechnology Past and Present
Leading to Science, Engineering, and Technology

Synthesis Lectures on Engineering, Science, and Technology

Each book in the series is written by a well known expert in the field. Most titles cover subjects such as professional development, education, and study skills, as well as basic introductory undergraduate material and other topics appropriate for a broader and less technical audience. In addition, the series includes several titles written on very specific topics not covered elsewhere in the Synthesis Digital Library.

Lying by Approximation: The Truth about Finite Element Analysis
Vincent C. Prantil, Christopher Papadopoulos, and Paul D. Gessler
2013

Simplified Models for Assessing Heat and Mass Transfer in Evaporative Towers
Alessandra De Angelis, Onorio Saro, Giulio Lorenzini, Stefano D'Elia, and Marco Medici
2013

The Engineering Design Challenge: A Creative Process
Charles W. Dolan
2013

The Making of Green Engineers: Sustainable Development and the Hybrid Imagination
Andrew Jamison
2013

Crafting Your Research Future: A Guide to Successful Master's and Ph.D. Degrees in Science & Engineering
Charles X. Ling and Qiang Yang
2012

Fundamentals of Engineering Economics and Decision Analysis
David L. Whitman and Ronald E. Terry
2012

A Little Book on Teaching: A Beginner's Guide for Educators of Engineering and Applied Science
Steven F. Barrett
2012

Engineering Thermodynamics and 21st Century Energy Problems: A Textbook Companion for Student Engagement
Donna Riley
2011

MATLAB for Engineering and the Life Sciences
Joseph V. Tranquillo
2011

Systems Engineering: Building Successful Systems
Howard Eisner
2011

To my parents who consistently encouraged me on the "scientific" path.

Nanotechnology Past and Present
Deb Newberry

ISBN: 978-3-031-00956-3 print
ISBN: 978-3-031-02084-1 ebook
ISBN: 978-3-031-00156-7 hardcover

DOI 10.1007/978-3-031-02084-1

A Publication in the Springer series
SYNTHESIS LECTURES ON ENGINEERING, SCIENCE, AND TECHNOLOGY Lecture #7

Series ISSN   2690-0300   Print   2690-0327   Electronic

Nanotechnology Past and Present

Leading to Science, Engineering, and Technology

Deb Newberry
Newberry Technology Associates

SYNTHESIS LECTURES ON ENGINEERING, SCIENCE, AND TECHNOLOGY #7

ABSTRACT

Nanoscience and nanotechnology, the application of the research-based nanoscale science, have changed significantly over the last three and a half decades. The "bucky" ball, 60 carbon atoms arranged like a soccer ball, and an often-used symbol of nanotechnology, was discovered in 1985 and 4 years later scientists at IBM were able to manipulate xenon atoms on a surface. In the intervening years, nanotechnology has evolved from a singly focused research topic to an understanding that infiltrates every aspect of science and engineering disciplines. In addition, nanotechnology, and both naturally occurring and engineered nanomaterials, have become the focus of legal, environmental, and application and regulation disciplines. The first portion of this text serves as an introduction to nanotechnology: the history, mathematical concepts, and instruments required to study and manipulate the world at the atomic scale. The later portion of the text discusses the connectivity of nanotechnology to the more traditional scientific disciplines as well as emerging technologies.

This text can serve as an introduction to the nanoscale for science, computer science, and engineering disciplines. It can also provide a valuable foundation for disciplines such as industrial hygiene, architecture, sociology, ethics, and the humanities. There does not exist an educational discipline, market segment, or career avenue which will not be impacted by nanotechnology.

KEYWORDS

nanoscience, nanotechnology, engineering, technology, societal aspects, science, atomic force microscope, undergraduate science, non-technical

Contents

Figure Credits

Figure 1.5: Based on: Miernicki, M., Hofmann, T., Eisenberger, I. et al. (2019). Legal and practical challenges in classifying nanomaterials according to regulatory definitions. *Nat. Nanotechnol.* 14, 208–216. Copyright © 2019, Springer Nature. Used with permission.

Figure 2.3: From: Vassil/Wikimedia Commons.

Figure 2.5: From: Auslender M. and Hava S. (2017). Single-crystal silicon: Electrical and optical properties. In: Kasap, S. and Capper, P. (eds), *Springer Handbook of Electronic and Photonic Materials*. Copyright © 2017, Springer International Publishing AG. Used with permission.

Figure 2.6: Based on: Mallakpour, S. and Soltanian, S. (2016). Surface functionalization of carbon nanotubes: fabrication and applications, *RSC Adv.* 6, 109916. Copyright © 2016, Royal Society of Chemistry. Used with permission.

Figure 3.2: Courtesy of Starpharma.

Figure 3.4: Reproduced by permission of Bruker and Hysitron.

Figure 3.5: Reproduced by permission of Bruker and Hysitron.

Figure 3.6: Image courtesy of Greg Haugstad, University of MN.

Figure 3.7: Reproduced by permission of Bruker and Hysitron

Figure 3.8: Based on: Inkson, B. J. (2016). 2 - Scanning electron microscopy (SEM) and transmission electron microscopy (TEM) for materials characterization. Hübschen, G., Altpeter, I., Tschuncky, R., and Herrmann, H. G. (2016). *Materials Characterization Using Nondestructive Evaluation (NDE) Methods*. Woodhead Publishing, pp. 17–43. Copyright © 2016 Elsevier Ltd. Used with permission.

Figure 3.9: Reproduced by permission of Bruker and Hysitron

Figure 3.10: Reproduced by permission of Bruker and Hysitron.

Figure 4.1: Roco, M. C. (2016). *Nanotechnology: Delivering on the Promise Volume 1, Building Foundational Knowledge and Infrastructure for Nanotechnology: 2000-2030.* © 2016, American Chemical Society. Used with permission.

Figure 4.3: Based on: Ajayan, P. M. (2019). The nano-revolution spawned by carbon. *Nature,* 575, 49–50. Copyright © 2019, Springer Nature. Used with permission.

CHAPTER 1

Introduction to Nanotechnology, a Basic Definition

1.1   THE WORLD AT THE MOLECULAR AND ATOMIC LEVEL

Human beings, in general, have always been inquisitive. We have wanted to investigate the world around us and the world inside of us. Historically, we were able to observe the world outside and away from us, for example the stars, much earlier than we have been able to observe the world within us. And just like observing the solar system, stars, galaxies, and universe has taken time and specific tools to be able to observe, measure, and understand—the world at the molecular and atomic level has taken time and special tools to understand. We have had to create specific tools and develop new rules of physics and chemistry in order to be able to understand what we observe at the molecular and atomic level—at the nanoscale.

Start with the basics. Figure 1.1 is a familiar picture of a carbon atom with the protons and neutrons in the nucleus and the electrons orbiting around the nucleus.

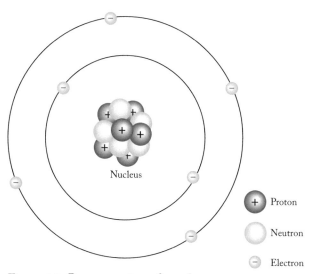

Protons are positively charged and electrons are negatively charged. The general rule, learned probably in the first or second grade, is that opposites attract and likes repel. In general, electrons being negatively charged will tend to repel away from each other and the protons being positively charged will also repel away from each other. Nuclear physics has identified subatomic particles, such as gluons, that explain why the nucleus, with all of the positively charged protons holds together. The physics of quantum mechanics can explain the orbital shape of the electron paths (shown here as circles for simplicity). In an atom or material the electrons and protons being opposite in charge will be attracted to each other.

Figure 1.1: Representation of a carbon atom.

This is the fundamental principle of nanotechnology—that likes repel and opposites attract. This will be a recurring theme throughout this text. It is easy to forget that molecules are charged entities with positive regions and negative regions, especially when we are studying complex structures such as proteins, amino acids, and cells. It is those regions of opposite charge that cause molecules to bind together into proteins, collagen, and eventually into cells in muscle tissue, brain tissue, and bones. Knowing this fundamental concept of "opposites attract and likes repel" allows an insight into the world at the molecular and atomic level.

1.2   CHEMICAL BONDING

One of the fundamental aspects of chemistry is the bonding of different atoms. There are two fundamental bonding types. These are ionic bonding and covalent bonding. Figure 1.2 shows the poster child for ionic bonding: the salt molecule.

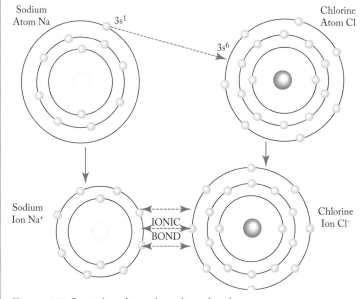

Figure 1.2: Ionic bonding: the salt molecule.

In the figure, the sodium atom on the left is represented by the nucleus of protons and neutrons with the electrons orbiting around the nucleus. The chlorine atom on the right is shown with the nucleus of neutrons and protons and the surrounding, orbiting electrons. The atoms are more energetically stable with eight electrons in the two outermost orbitals for these atoms with three electron orbitals (the furthest away from the nucleus) and two electrons in the innermost orbital. The two electrons that are closest to the positively charged nucleus due to the protons do not usually contribute to chemical bonding. That is, the force binding them to the nucleus is much stronger than most attractions to other nuclei or atoms. It is only the electrons that are further away from the nucleus that have a lower attractive force to the nucleus that will contribute to the chemical bonding between atoms and larger structures such as molecules.

Consider now what is known as a salt molecule with one sodium atom and one chlorine atom. The sodium atom has one electron in its outermost shell. Note that the number of electrons in each orbital is defined by physics interactions. The desired number from a chemistry/physics standpoint is eight. The chlorine atom has seven electrons in its outermost shell or orbital. When a sodium atom and a chlorine atom come close together the outermost electron in the sodium atom jumps over to the chlorine atom and completes its outermost shell of eight electrons.

The resulting salt molecule with a sodium and a chloride atom is configured with the sodium atom missing one electron so it has a net positive charge because it has one more proton in the nucleus than it has electrons orbiting around that nucleus. The chlorine atom because it has acquired one additional electron has a net negative charge. In general, the molecule, salt, consists of a sodium atom and a chlorine atom, which are now ions, and it has a net region of positive charge closer to the sodium atom and a region of net negative charge closer to the chlorine atom. If we bring more salt molecules into play, opposites attract and likes repel and the result is a salt crystal, as shown in Figure 1.3.

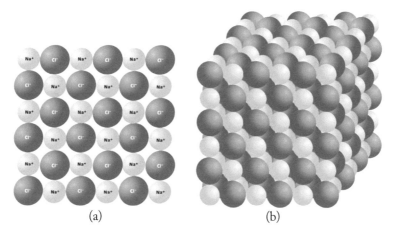

(a)   (b)

Figure 1.3: Salt crystal viewed in a plane (a) and as a cubic crystal (b).

Each edge of that cubic salt crystal will have a checkboard arrangement of positive and negative regions.

The other major type of chemical bonding is covalent bonding. The poster child for covalent bonding is the water molecule shown in Figure 1.4.

The water molecule, H_2O, is made up of one oxygen atom and two hydrogen atoms. The hydrogen atom, the simplest atom, has one proton in the nucleus and one electron orbiting around that nucleus. Within an atom the inner layer of electrons closest to the nucleus is most energetically stable when it has two electrons in that orbit.

The oxygen atom with its atomic number of 8 has eight protons in the nucleus and eight electrons orbiting around that nucleus. Two of those electrons are in the orbit closest to the nucleus and the remaining six electrons are orbiting a little bit further away from the nucleus. From an energy standpoint, the oxygen atom would be more stable with eight electrons in its outermost shell, but then it would be neon. Each hydrogen atom has one electron orbiting around the single proton nucleus. For each of these hydrogen atoms, the electrons, being negatively charged, will experience a stronger force of attraction to the 8 positively charged protons in the oxygen nucleus than the attractive force to the one proton in the hydrogen nucleus. As a result, the hydrogen atom electrons will spend a greater portion of their orbital "time" closer to the oxygen atom. This will create an outermost layer of the desired eight electrons. This is shown in the upper-left-hand portion of the figure.

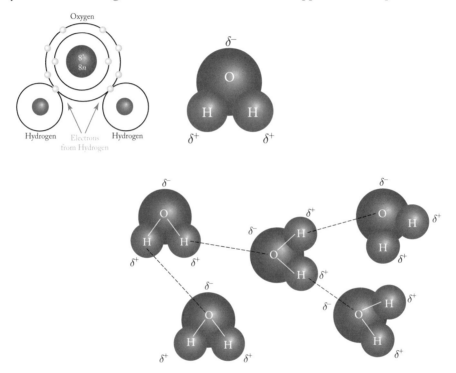

Figure 1.4: Water molecule(s) and covalent bonding.

The oxygen atom has a positive eight protons in its nucleus and so the attraction between the electrons in the hydrogen atom and that eight-proton positive charge in the nucleus of the oxygen atom is greater than the attraction to the single proton in the hydrogen nucleus. This is covalent bonding or sharing of electrons.

The two hydrogen atoms will "share" their two individual electrons with the oxygen atom resulting in a water molecule. Again, similar to the salt molecule, there is a nonuniform charge distribution in the water molecule.

Observing the water molecule from above, from the top of the oxygen, most of the time a negative charge is observed because of the eight electrons that are orbiting in the outermost orbital of the oxygen. Observing the water molecule from the bottom or from the hydrogen side, as it is seen in Figure 1.4, most of the time those electrons which are associated with the hydrogen atom are going to be associated more with the oxygen atom and what is observed from the bottom of the water molecule is the positive charge of the hydrogen nuclei or the protons. Observing the water molecule from one direction, it appears to carry a negative charge most of the time and observing that same molecule from a different direction, a positive charge is observed. This difference in charge distribution is represented by the Greek small letter, delta (δ), with either a plus or a minus superscript. This designation is used in Figure 1.4.

The water molecule is the poster child for covalent bonding with sharing of electrons.

There is not a uniform charge distribution in either case: both the salt molecule with ionic bonding and the water molecule with covalent bonding have nonuniform charge distribution. This nonuniform charge distribution is the reason various materials have different properties; proteins fold the way they do and antibodies can attach to "invaders."

1.3   SENSE OF SCALE MEASUREMENTS, CALCULATIONS, AND DEFINITIONS

In some cases, it is easy to get a sense of the largeness of space by looking up at the sky at night and understanding that the earth and its population represent a very small piece of a very large universe. Changing direction and heading down to the very small can pose more of a challenge in terms of understanding the world at that scale.

The prefix *nano* is truly just that, a prefix. Very much like *centi* is the prefix for centimeter or centi means 100th of something. Hence, a centimeter is 100th of a meter. The prefix *nano* means one billionth or 10 to the -9 of an entity so 1 nanometer is one billionth of a meter. The prefix *milli* means 1,000th and a millimeter is 1,000th of a meter in terms of the length scale. Prefixes like *centi*, *milli*, *nano*, *pico*, *femto*, and all the other prefixes can apply to entities other than length. Table 1.1 lists the various prefixes and their power of ten designation.

For example, we can have a nanosecond; that would be one billionth of a second. A nano-gallon represents one billionth of a gallon or a centigallon would be 100th of a gallon. A nanomile is one billionth of a mile. The prefixes can be applied to any measurement.

In Table 1.1, the exponential notation that accompanies that prefix is also listed. For example, nano is one billionth so that would be 10 to the -9. A millimeter would be 10 to the -3 because it's one 1,000th of a meter. Centi which means 100 would be 10 to the -2 and so on. The table includes multiple prefixes that apply to different exponential notations. Remember these prefixes can apply to any unit of measurement. A centisecond is 100th of a second and a picosecond would then be 10 to the -12 of the second, a very short period of time.

For the purposes of this text, nano will predominantly apply to the length scale, and the discussion will involve nanometers and atomic dimensions. When investigating nanotechnology, various calculations will need to be performed and it is useful to have an understanding of the rules of exponents. The rules of exponents are shown in Table 1.2.

Table 1.1: A list of metric prefixes, symbols, and exponential notation			
Prefix	**Symbol**	**Multiplier**	**Exponential**
Yotta	Y	1,000,000,000,000,000,000,000,000	10^{24}
Zetta	Z	1,000,000,000,000,000,000,000	10^{21}
Exa	E	1,000,000,000,000,000,000	10^{18}
Peta	P	1,000,000,000,000,000	10^{15}
Tera	T	1,000,000,000,000	10^{12}
Giga	G	1,000,000,000	10^{9}
Mega	M	1,000,000	10^{6}
Kilo	k	1,000	10^{3}
Hecto	h	100	10^{2}
Deca	da	10	10^{1}
No Prefix Means:		1	10^{0}
Deci	d	0.1	10^{-1}
Centi	c	0.01	10^{-2}
Milli	m	0.001	10^{-3}
Micro	μ	0.00001	10^{-6}
Nano	n	0.000001	10^{-9}
Pico	p	0.0000001	10^{-12}
Femto	f	0.00000001	10^{-15}
Atto	a	0.000000001	10^{-18}
Zepto	z	0.0000000001	10^{-21}
Yocto	y	0.00000000001	10^{-24}

Table 1.2: Rules of exponents
$(10^a)^b = (10)^{ab}$
$(10^a)(10^b) = 10^{a+b}$
$10^a/10^b = 10^{a-b}$
$1/10^a = 10^{-a}$

It is important to ensure an understanding of these rules of exponents. It is also important that special attention is paid to the sign on the exponent, whether it's negative or positive. There are multiple worksheets available on the Internet to practice the application of these rules of exponents.

In general, when nanotechnology is discussed the prefix nano is applied to a scale of length, with nanometer, abbreviated nm, as the standard. To get a sense of the size of a nanometer consider the fact that a hydrogen atom is approximately 1/10 of a nanometer in width. Lining up 10 hydrogen atoms in a row would equal 1 nm. So an atom will have dimensions of less than a nanometer. However, molecules with multiple atoms will have sizes much larger than a nanometer. A DNA strand is approximately 2–3 nm in diameter and can be many nanometers or centimeters long.

In the U.S., and also in other countries, it was important that "nanotechnology" be defined. The definition of nanotechnology would often determine which category may be appropriate for any given research. Defining the research category is critical to defining which agency or institution is appropriate to financially support such research. The categories could range from medical diagnostics to plant fertilizers to coatings for military vehicles.

Historically, nanotechnology within the U.S. has been defined as any material or entity created that had one of its dimensions in the 100 nanometer range. For example, from the Department of Defense or the National Science Foundation, organizations that fund a large amount of research, a project would fall into the nanotechnology category if any one of the three dimensions X, Y, or Z was equal to or less than 100 nm. A researcher could be investigating graphene or carbon nanotubes (CNTs). CNTs might be millimeters long, that is, thousands of nanometers in length, but only 5 nm or 50 nm in diameter. That research dealing with those CNTs would fall into the "nanotechnology" category even though they were millimeters long because they have one dimension that was less than 100 nm. That definition continues to be the guiding definition for research funded in the U.S.

The definition of nanotechnology can get extremely complicated when we enter the world of Environment, Health, and Safety (EHS). This area involves regulatory agencies that are in the process of trying to test, quantify, and define health hazards or lack of health hazards, environmental and safety impacts of any and all materials. Defining nanotechnology from an environment, health, and safety standard standpoint can become complicated. In the U.S. regulatory agencies continue to debate exactly how to quantify nanomaterials or EMs (engineered nanomaterials) or engineered nanoparticles from an environment, health, and safety standpoint. Figure 1.5 is from an article in the March 2019 issue of *Nature Nanotechnology*. This particular figure represents the definition of nanotechnology as defined in the European Union (EU).

As seen in Figure 1.5, different organizations even within the same set of countries like the EU will define nanotechnology differently.

In the cosmetic industry, nanotechnology or nanomaterials are defined as materials with one or more external dimensions or an internal structure on the scale of 1–100 nm. Therefore, working with the cosmetic industry in the EU, if the cosmetic has particles of collagen, proteins, or oils, for example, that are larger than 100 nm then that cosmetic does not fall into the nanotechnology category. However, if the cosmetic does indeed contain particles or have material with an internal structure between 1 and 100 nm, that cosmetic falls into the nanotechnology category and is subject to the restrictions of nanotechnology regulation. The cosmetic industry has used encapsulated materials, such a vitamin E, collagen, and other cell nutrients, contained in capsules small enough to penetrate the outer dead layers of skin. Although the enclosing capsule may not fall under the nanoscale definition, the contained materials were at the nanoscale.

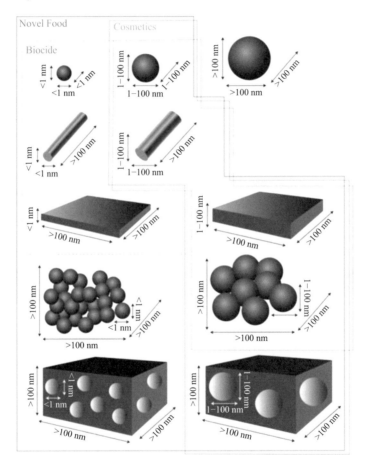

Figure 1.5: Depiction of various definitions of "nanotechnology."

With regard to food regulation, again in the EU, a food falls into the nanotechnology regime if it contains any nanomaterial that was intentionally produced and has one or more dimensions of the order of 100 nm or less or if it's composed of discrete functional parts that have one or more dimensions of the order of 100 nm or less including structures, agglomerates, or aggregates. This means, from a food standpoint, if an entity is composed of material that is less than 100 nm in size it will be deemed nanotechnology, but the definition is also expanded to include aggregates and properties that are characteristic of what occurs at the nanoscale.

These definitions are important because one of the attributes of nanomaterials is that physical and material properties will change as the size is diminished. For example, gold in large quantities, like a coin, will have a specific melting temperature. However, gold nanoparticles or gold that is much smaller than the size of a coin will have a different melting point. There is also a substantial dependence of the properties of nanoscale material based on the manufacturing, measurement, or application environment. These environmental aspects can include temperature, humidity and pressure—all of which can modify the property of the nanoscale material. Another constraint of the nanoscale material involves the configuration of the nanomaterial. Shape and purity are two aspects that fall into this category. The shape of the nanomaterial will significantly impact not only the physical properties but also the material interaction with the external environment.

In part, this change in material properties is due to the fact that the ratio between surface area and volume is modified significantly as the material is divided into smaller and smaller units. An example of this is shown in Figure 1.6 where we have a cube of a given dimension as shown in the figure.

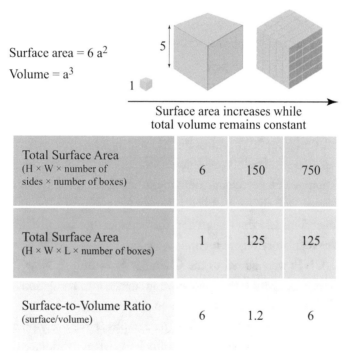

Surface area $= 6\,a^2$

Volume $= a^3$

Surface area increases while total volume remains constant

	6	150	750
Total Surface Area (H × W × number of sides × number of boxes)	6	150	750
Total Surface Area (H × W × L × number of boxes)	1	125	125
Surface-to-Volume Ratio (surface/volume)	6	1.2	6

Figure 1.6: Comparison of surface area for a constant volume broken into smaller cubes.

A cube will have a surface area of six times the area of each one of the sides. If the volume is kept constant but the cube is broken into smaller cubes, the surface area exposed for the same volume of material goes up dramatically.

Chemical reactions are dependent upon exposed surface area. Therefore, the large cube broken into smaller pieces, resulting in a much greater surface area, will have a much higher chemical reactivity than the single larger cube of material. Because it is the exact same volume of material in both cases it is implied that both cost and mass are the same. Although it is acknowledged that there is a cost associated with "sawing" or dividing the larger cube into smaller pieces.

All material properties such as boiling point, tensile strength, chemical reactivity, and the ability to dissolve in water will change depending upon the surface area of the material. For example, when dissolving sugar in water or salt in water, it will be much faster to dissolve granulated salt or granulated sugar than the time it would take dissolve a large crystal of sugar or salt in the water.

1.4   AN ISSUE WITH DEFINITIONS

Again, looking at Figure 1.5, from a biotechnology or biological regulation standpoint, nanomaterial is defined as a "natural or manufactured active substance containing particles in an unbound state or as an aggregate or agglomerate where 50% or more of the particles in the number size distribution has one or more external dimensions in the size range of 1 nm–100 nm." That is a quote from the article in *Nature Nanotechnology* referenced in Figure 1.5. From this standpoint, the regulation is dependent on the percentage (50%) of the aggregate material with dimensions within the 1 nm and 100 nm range. This is a case where the composition and the makeup of the material assessed needs to be understood. Based on this definition, if you had a beaker of material but only 10% of it was smaller than 100 nm it would not fall under this particular regulatory control. Finally, there is a recommendation from 2011 within the EU which defines a nanomaterial as any natural or incidental or manufactured material containing nanoparticles in the size range of 1 and 100 nm, independent of the composition percentage or amount. However, there are some materials such as graphene flakes, fullerenes, and single-walled CNTs that are going to fall under this category even though they have dimensions less than 1 nm and therefore are not in the defined size range. They will still fall under this regulatory requirement.

Later in 2019, CNTs were added to the ChemSec Substitute It Now (SIN) list. ChemSec is a Swedish non-profit organization that may have influence into regulations within the EU and other countries. To date, no action has been taken on this item by the EU.

The information discussed in the previous paragraphs is relatively new and illustrates the complexity of defining nanomaterials or nanotechnology when working with any small-scale material. The importance is that these nanomaterials are going to fall under regulatory aspects involving health, environment, and safety. In order to define the regulation, it is necessary to define a nanoscale material. In the U.S., the definition continues to evolve as different government agencies wrestle with the definition of a nanomaterial. Once a nanomaterial is defined, most likely by dimensions, how that material is tested and measured will determine how that material will be regulated considering environment, health, and safety aspects as well as the manufacturing and measurement environment and the intended application.

1.5   SCIENCE AND TECHNOLOGY

A word about the definition of science and technology. Often the term nanoscience will be used interchangeably with the term nanotechnology. Sometimes nanotechnology is used and nanoscience is implied. Historically, the word "science" applies to research or undefined aspects of a subject. A researcher observing nanomaterials, creating nanomaterials, or investigating nanomaterials for an undefined purpose, would be doing nanoscience. Nanotechnology is the definition that engineering uses where the science that has been determined is applied to some sort of an application. Nanotechnology would be the application of nanoscience. For example, recently researchers have been able to create silver nanoparticles that have a core of about 10 atoms of silver; this core is encased in additional silver atoms as well as additional constituents. This particular nanoparticle is certainly in the nanoscale range and it has the potential for being able to treat certain diseases. The researcher that created this nanoparticle developed a process to create these nanoparticles that have specific composition and also specific properties is doing nanoscience. However, when a medical student, doctor, technician, or biotechnologist takes these nanoparticles and begins to investigate the potential application of these nanoparticles to impact disease molecules, then their work could be referred to as nanotechnology.

In many cases the terms *nanoscience* and *nanotechnology* are used interchangeably without a substantial distinction between science and technology. In general, nanotechnology applies to both the research and the application with the term nanoscience being used less and less.

CHAPTER 2

The History of Nanoscience and Nanotechnology

2.1 INTRODUCTION

The desire to create materials and products with certain properties that perform specific functions has been a part of human nature since the beginning of time. In many cases, properties or functions were found by accident or trial and error. For example, think of the development of Scotchgard when a research assistant dropped a solution on his tennis shoe by accident and it repelled water and soiling. The researchers immediately realized the potential product that would evolve. Sometimes need can be the driving force such as happened with Post-it notes. They were developed because a choir director needed to attach notes to his score.

The same is true for developments and products that involve nanotechnology. At the time of development, the contribution of nanoscale materials was not understood or even acknowledged. However, as new tools have been developed that allow us to observe the world at that scale, an understanding of the nanoscale involvement is being realized.

2.2 THE BEGINNING OF THE "MODERN" ERA OF NANOTECHNOLOGY

The modern era of nanotechnology is often believed to have started in 1959 when Richard Feynman, a physicist and Nobel Laureate, gave a speech at Caltech titled "Plenty of Room at the Bottom." He was speaking to a group of physicists at one of the international physics meetings and in his speech, he challenged physicists to think about things at the atomic level. He asked the physicists to think about what it would take to be able to observe and manipulate at the atomic level or the nanoscale.[1]

After that talk, Caltech researchers began to take Richard Feynman's challenge seriously and started thinking about how to create, manipulate, observe, design, and measure at the nanoscale. In fact, one of the challenges that Dr. Feynman gave the physicists was to write the entire content of the Library of Congress on the head of a pin. Perhaps this thought experiment gives you an idea of just how small the nanoscale is.

---

[1] Toumey, C. (2009). "Plenty of Room" revisited. Who was Richard Feynman and what did he actually say about nanotechnology? *Nature Nanotechnology*, 4(12) p. 781. DOI: 10.1038/nnano.2009.356.

Within the next few decades companies such as Hewlett-Packard, Motorola, IBM, and General Electric started looking at the nanoscale, and scientists at those companies began creating the tools that we now use to observe the world at the nanoscale. These tools included scanning tunneling microscopes (STM), atomic force microscopes (AFM), and transmission electron microscopes (TEM). These tools will be discussed in greater detail in the next chapter. Some of these advances are shown in Figure 2.1.

1959
Feynman gives an after-dinner talk describing molecular machines building with atomic precision

1974
Taniguchi uses the term "nano-technology" in a paper on ion-sputtr machining

1981
STM invented

1986
AFM invented

1989
IBM logo spelled in individual atoms

1997
First company founded: Zyvex

Figure 2.1: Some major developments in the recent history of nanotechnology.

2.3   THE EARLY APPLICATIONS: TRIAL AND ERROR

However, humanity has been using attributes of nanoscale materials for centuries without knowing what was happening at the nanoscale. Ancient Egyptians used nanomaterials or created nanomaterials that resulted in an enduring century-old shine on some of their statues. Artists in Europe used nanoscale gold and silver particles to create the different colors in the beautiful stained glass windows that we find in the cathedrals in Europe (Figure 2.2).

Figure 2.2: Stained-glass window from a European cathedral.

It was through trial and error that the artisans learned that when you mixed different elements, predominantly gold, silver, and copper that had been manipulated for a certain period of time, in with the molten glass that you can create red glass from gold or green glass from copper or silver nanoparticles. They did not understand that they were dealing at the nanoscale, at the atomic level or with very small particles of gold and silver. Through trial and error, a method was developed that resulted in the creation of beautiful colors in those windows.

Finally, a classic example of the application of nanotechnology in ancient times is a goblet known as the Lycurgus cup (Figure 2.3). This goblet was made of glass and had a unique property that when the goblet was held in front of a light with the light shining onto the glass it appeared green. The goblet was reflecting green light. However, when a light was placed inside the goblet, like a candle, the goblet would transmit red light and the goblet would appear red. This color difference depending on the location of the light source is shown in the figure. On the left-hand side is the color that the goblet appears in reflected light and on the right-hand side is the color the goblet appears when illuminated from the inside. This goblet is thousands of years old and has resided in the British Museum of History for a long time. Only recently researchers were able to study the cup and understand why it had such unique properties. The researchers learned that this goblet contains nanoparticles of both silver and gold that are about 50–70 nm in diameter. Varying sizes of the nanoparticles interact with the spectrum of visible light differently. Also, the atomic structure of gold and silver is different, and each element will interact with light differently. Light that is reflected from the Lycurgus cup is in the green wavelength which is around 540 nm in wavelength, the transmitted light through the glass has a wavelength of around 700 nm in wavelength, which makes the glass appear to be red.

Figure 2.3: The Lycurgus Cup was determined to contain nanoparticles of gold and silver. The image on the left shows the reflected green color and the one on the right shows the transmitted wavelength of red light.

More recently, researchers, engineers, and scientists have been using nanotechnology and applying nanoparticle characteristics to different materials without necessarily realizing that they were working at the nanoscale. Again, even for the modern-day application of what we now know as nanotechnology, many processes were developed by trial and error.

One example of this trial and error process is the growing of a large single silicon crystal called a boule. This silicon crystal is the foundation for the creation of semiconductor devices. To create transistors and semiconductor devices it is necessary to start with a very pure material. That is, a pure crystal of silicon as the base material for most electronics is required. The silicon crystal must have as few defects as possible in order to ensure the accurate operation of the transistors and to ensure the highest probability of success in the transistor fabrication process.

The method that is used to create this single crystal of silicon, which can end up being 400 lb, about 4 ft long, and 12–15 in. in diameter, is grown out of a molten bath of silica as shown in Figure 2.4.

Starting Crystal and Pull Rod   Silicon Crystal Forming

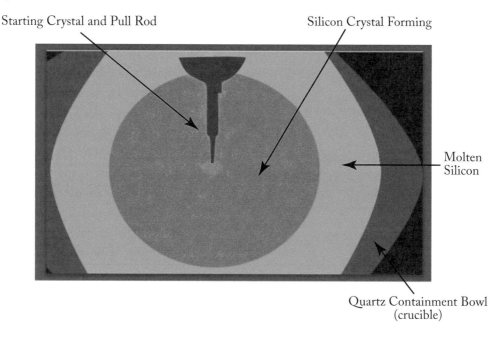

Molten
Silicon

Quartz Containment Bowl
(crucible)

Figure 2.4: Drawing (top view) of a single crystal of silicon being pulled from a molten pool of silicon.

Silicon, which is basically sand as pure as possible, is heated up and melted in a quartz bowl. A single crystal of silicon is attached to a thread and dipped into the molten silicon. That single crystal will attract other crystals of silicon in an organized structure. That thread is just slowly rotated and pulled out of the molten bath of silicon. As the crystal is pulled out of the molten bath, more and more silicon atoms attach to this crystal structure and the result is a very large, pure, single crystal of silicon. The crystal structure of silicon is shown in Figure 2.5. This structure is very similar to that of a diamond and germanium.

It took many years of trial and error to determine the temperature for the molten silicon, the rate at which to remove or pull the string out of the molten silicon, and how fast that that string or that thread with the starting crystal was to spin. All three of those characteristics that define the success of growing a single crystal of silicon were evaluated, established, and defined by trial and error. It has only been recently with the advent of advanced equipment, such as atomic force microscopes and scanning electron microscopes, that the ability to understand, at the atomic level, why the recipe for creating a single crystal of silicon at a specific temperature, rate of pulling, and rate of turning was successful. The insights that have been gained using the tools of the nanoscale have resulted in improvements in the purity and quality of the silicon boule and also allowed an increase in the diameter of the boule from 3 cm to the current diameter of 30 cm.

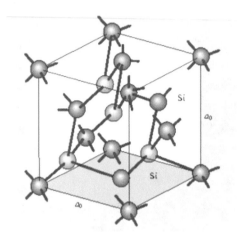

Figure 2.5: The atomic levels crystal structure of silicon.

Another example of trial and error which impacted the nanoscale is in the hardening of aluminum. Often the process of using alloys, heating, quenching, annealing, or age hardening for various metals was discovered "by accident" or trial and error. Aluminum was of special interest, first, because it is the third most abundant element behind silicon and oxygen. Second, it is very lightweight, which is one reason that aluminum was used in early armor and is often used as protective plating. However, aluminum was a very soft metal, therefore it needed to be combined with other compatible metals like copper and magnesium. The alloying of materials such as aluminum involved the mixing of the metals at high temperatures then a rapid cooling method such as water quenching. In 1905, A. Wilm, a German metallurgist working with aluminum and alloys, was using this process of heating and then rapidly cooling to improve the strength of the alloyed material. After returning to work from the weekend he found that the alloy was now significantly stronger than it has been two days earlier. This introduced the process known as "age hardening." This three-step process resulted in an aluminum alloy that had many of the desired physical properties.

Another approach used for forming metals into desired shapes was to heat the metal or alloy up to a certain temperature where the material was malleable but not molten and force the material through a form or template to create a beam of a certain shape or diameter. It was found that during this process, often the resulting product, after cooling, was stronger than the starting material. Through a series of trial and error experiments, the proper combination of alloy concentrations, temperature, and force or pressure was determined which resulted in the desired properties.

It was trial and error and a lot of experiments that ended up resulting in the metals that we have today with the properties that are needed for bridges, roads, buildings, homes, and all types of construction. Today, because we have atomic force microscopes, nanomechanical test equipment, and scanning electron microscopes not only can the physical properties be measured but the atomic structure can be observed, evaluated, and modified to enhance the desired properties. Secondary benefits which occur as a result of understanding the atomic level structure of these materials are that alternate materials might be used which could impact cost, weight, manufacturability, and quality.

Because of the development of these new tools which allow the observation and measurement at the nanoscale we can understand why various materials have the properties that they do at the macroscale, microscale, and nanoscale. These tools also allow insight and understanding into the effectiveness of one process over another. The understanding of the manipulation which is part of the trial and error process is now available.

The converse is also true. An understanding of the atomic level structure of certain materials, for example hardened aluminum, leads to an understanding of the relationship between the atomic structure and specific properties. That knowledge can be used, as manufacturing capabilities reach the nanoscale, to create materials with specific atomic structures that have the desired properties.

An example of this type of inverse engineering is using CNTs and/or individual sheets of carbon atoms: graphene. It has been found that graphene may have different properties depending upon the number of layers within that sheet. By combining CNTs with specific properties with sheets of graphene, researchers even have been able to create a phone which includes a graphene screen and battery as well as having the potential for graphene and CNT antennas and electronics. This phone has thermal and structural characteristics that are desired but it also may be bendable and very flexible. This "graphene phone" may have multiple applications in space, construction, and embedded electronics.

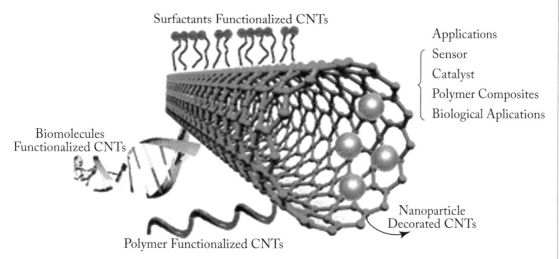

Figure 2.6: Carbon nanotube (CNT) with various functionalization possibilities shown.

The same is true regarding CNTs. Researchers, scientists, and engineers can now create CNTs that are single-walled or multiwalled. The multiwalled CNTs have tubes within tubes. CNTs, similar to graphene sheets, can have different properties dependent on many factors. Individual atoms or molecules can also be attached to CNTs increasing the library of properties from the material field into the biological arena (Figure 2.6). The figure includes potential applications, including biological applications.

In the biological area, the understanding of diseases, prevention, detection, and treatment or cure has been based on multiple years of research often requiring millions of dollars followed by lengthy and complicated clinical trials. In recent years, the understanding of biological systems has grown substantially. In many cases this growth is due to the observation and measurement capabilities available at the nanoscale. Consider the series of images shown in Figure 2.7.

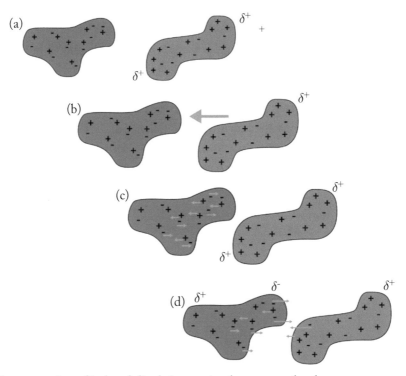

Figure 2.7: Representation of induced dipole interaction between molecules.

Figure 2.7(a) shows a representation of two molecules. The molecule on the left, shown in blue, has a uniform charge distribution whereas the one on the right is a dipole molecule with two regions that have a positive charge shown by the delta symbol with a plus superscript. When the two molecules are moved closer together , as shown in Figure 2.7(b), the atomic arrangement and/or the arrangement of the electrons within the atoms may be affected. This affect is shown on Figure 2.7(c) as the positive entities in the blue, left molecule are repelled by the positive charge in the right molecule. The lower image, Figure 2.7(d), shows the dipole that has been induced in the left molecule because of the movement of the repelled positive charges in the blue molecule and the remaining negative charges in the right portion of that blue molecule. The induced dipole is represented by the delta symbols with the positive and negative superscripts.

This example of an induced dipole is the path often used by drug molecules to become attached to the various portion of a protein or enzyme molecule. Visibility into the nanoscale has allowed for more refined and accurate observation and measurement of the atomic structure and charge strength of both biological entities such as proteins and developed drug molecules. This knowledge has led to fewer drug variations that need to be tested in clinical trials which, in turn, results in shorter and less expensive trials. This knowledge has also resulted in the development of drugs with the potential of fewer side effects and better efficacy.

2.4 CURRENT EXPECTATIONS

Understanding of the world at the nanoscale is impacting every educational discipline, multiple market segments, and almost every government agency. The widescale interest, research, and investment in nanoscale phenomena has led to high expectations for the technology, applications, and ability to address world problems. These problems include biological threats such as diseases, bacteria and fungal infections, hunger and drought situations, environmental concerns with regard to water and air quality, exploitation of resources, and military capabilities. The expectation that nanotechnology may be able to address and solve these problems is balanced by the concern regarding unintended or unknown consequences of engineered nanomaterials or nanomaterials that may occur naturally.

Many of the positive expectations and concerns regarding nanotechnology focus on the technical aspects of this science. Developing drugs to treat diseases, filters to help clean water and air, creating manufacturing processes that are "green," researching and defining end results of the use of nanotechnology, etc. are all based on technical aspects. There is also a less tangible aspect of nanotechnology which involves societal aspects that deal with human nature and human beings. The expectations, fears, and concerns of a society are often not founded on solid facts or valid scientific knowledge but can be based on or influenced by rumors or hearsay, incorrect explanations, or lack of understanding of the total situation. In some cases, thoughts, decisions, and positions may be totally driven by political or financial scenarios. Hence, expectations can be varied and uncertain.

A thought experiment: It is well known that the Ganges River in India is extremely polluted with a myriad of pollutants, from waste material to bacteria to dirt and grime. It is also known that the Ganges River is the only source of water for thousands of people. Suppose a company could create a water filter, perhaps the size of a piece of paper. Suppose also that once water direct from the Ganges River was poured through this filter that clean, pure water was the result—no dirt or bacteria or parasites or grime. Suppose also that the company could create these filters for 25 cents each and decided to sell them in India. With the population of India and the fact that many of the water sources are significantly polluted, not just the Ganges River, there would be an opportunity to make a great deal of money. These filters could be sold worldwide. However, it is noted that the people in India who get their water from these rivers may not have 25 cents to buy these filters, even if they are reusable for a while. What is your obligation?

CHAPTER 3

Tools Used in Nanoscience

3.1 INTRODUCTION

Optical microscopes have evolved and been transformed into tools that can create images of individual atoms and, in some cases, the bonds between them. The developed tools of nanoscience and nanotechnology are reviewed in this chapter. Topics include (1) Evolution from Eyes to Microscopes, (2) Optical Microscopes and Beyond, and (3) "Scopes for Today".

3.2 EVOLUTION: FROM EYES TO MICROSCOPES

In early history we learned about the world around us by using the five senses. We understood the properties and aspects of materials and living things based on what we could see, hear, touch, taste, and smell. The world at the microscale could not be understood based on our senses and many misconceptions arose. At one point it was believed that living things, like fleas, were created from inanimate objects like a suit of armor which had been worn for a long time. The connection between fly eggs, larva, and adult insects was not understood until the 15th century.

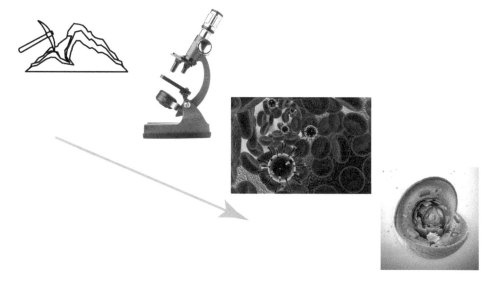

Figure 3.1: Methods and ability to observe the world around us.

In the 13th century magnifying lenses were created which allowed the sense of sight to be enhanced and smaller objects could be studied. Even though there is significant debate about who exactly "discovered" or created the first optical microscope, the name most commonly acknowledged is Antonie van Leeuwenhoek (1632–1724) who opened the world of biology with his newly developed instrument. Earlier, in 1610, it is said that Galileo Galilei reversed his telescope and with a few modifications was able to magnify small objects.

Independent of who may have actually created the first microscope, the availability of such a tool had a significant impact on our understanding of the world around us. For several centuries it remained the primary tool for studying flora and fauna, biological entities such as cells, tissues, and solid materials such as bone. There are limitations of the optical microscope which can be explained by a general "rule of thumb."

This rule of thumb states that the size of an object that you can image (or create) is approximately equal to ½ of the value of the wavelength used to illuminate (or form) it. Note: the application of this rule of thumb as it applies to the fabrication of electronic devices will be covered in Chapter 6.

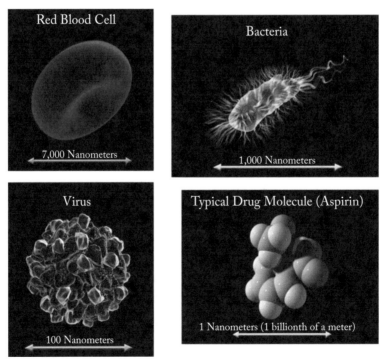

Figure 3.2: **Relative sizes of the world of biology, courtesy of Starpharma.**

For the application to the optical microscope the rule says that the size of an object that we can "see" is equal to ½ of the wavelength of the light used to illuminate our sample. Light in the visible range of the electromagnetic spectrum has wavelengths in the range of approximately 400 nm (blue) to 700 nm (red). Therefore, application of this rule states that the smallest object that should visible with an optical microscope, assuming it was illuminated with "white," visible light would be 200 nm in size. Figure 3.2 shows the size scale of several biological entities ranging from the large red blood cell with a diameter of 7,000 nm to an aspirin molecule with a diameter of approximately 1 nm.

Therefore, with an optical microscope a red blood cell could be easily observed and a bacteria could be seen but smaller entities such as a virus or an aspirin molecule could not be seen. Over time, microscopes have continued to improve and illumination with shorter wavelengths of light such as lasers are being used. This allows for the observation of smaller objects.

3.3   OPTICAL MICROSCOPES AND BEYOND

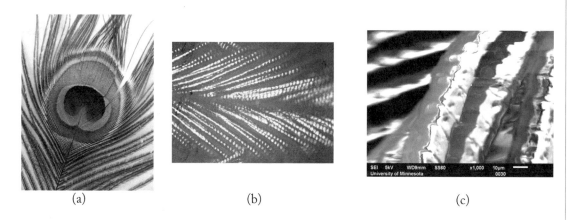

(a)   (b)   (c)

Figure 3.3: (a) Peacock feather observed with no magnification, (b)optical microscope at 100×, and (c) with a scanning electron microscope.

Figure 3.3 shows a progression of images of a peacock feather: (a) is what can be observed with our eyes, (b) is the barbules (substructure) of the feather when observed with an optical microscope at 500X, and (c) is the substructure when observed with a scanning electron microscope. With the advent of the more powerful scopes that enable observations at the nanoscale, scientists have learned that in nature much of the iridescent colors, such as those found in a peacock feather, the morpho blue butterfly, and hummingbirds, are due to structure and not a chemical pigment. The structure of the melanin barbules in the peacock feather can be seen in the microscope images.

3.4 "SCOPES" FOR TODAY

One of the most commonly used tools of nanoscience is the atomic force microscope (AFM) shown on Figure 3.4.

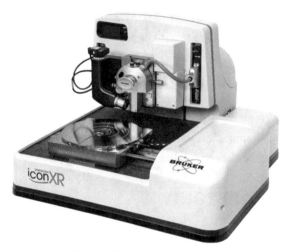

Figure 3.4: Atomic force microscope.

The instrument shown in Figure 3.4 is a research-grade tool and has very high resolution capability. In recent years, several companies have developed simpler, bench-top AFMs that are now found in many community colleges and high schools. The AFM operates by placing the material under study on a movable plate, shown in the lower right-hand section of the instrument, and moving that material up into contact with a cantilever beam. The cantilever has a very sharp tip on it that moves over the material surface. The tip interacts with the electron field around the atoms in the material. A laser is reflected off of the top of the cantilever and tracks the movement of the cantilever up and down as it is scanned/moved across the surface. Computers are then used to create an image from the laser information. The operation of the AFM and an image of the cantilever tip are shown in Figure 3.5.

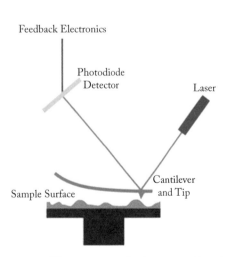
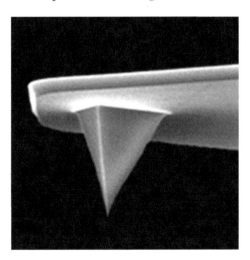

Figure 3.5: The operational schematic of an Atomic Force Microscope (AFM) on the left and a scanning electron microscope image of the tip at the end of the cantilever on the right.

Figure 3.6: **AFM image of water nanodroplets (largest ~300 nm in diameter) condensed on hydrophobized glass surface. Image courtesy of Greg Haugstad, University of MN.**

Another commonly used tool of nanoscience is the scanning electron microscope (SEM). This tool was created and used as a research vehicle as early as 1942 and continues to be improved to create images of finer and finer resolution. It is no longer the large, special-purpose unit that it was initially, and now can be found in many college labs as a tabletop model. Figure 3.7 is a picture of the JOEL IT500HR SEM, which is a research-grade scanning electron microscope.

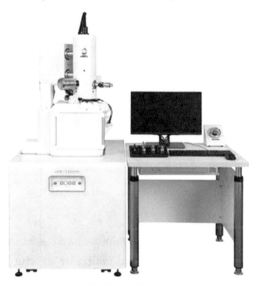

Figure 3.7: **JOEL IT500HR SEM equipment.**

As the name implies, the scanning electron microscope uses electrons, rather than visible light to study a surface. The energy of the electron is related to a wavelength. The higher the energy of the electron, the shorter the wavelength, and, based on the rule of thumb, the smaller the objects that can be imaged. The electrons are collimated in a small beam that is rastered over the surface under study. The SEM operates in a vacuum environment. Although modifications have been made which allow an environmental chamber to be introduced which can be temperature and humidity controlled. These tools are called environmental scanning electron microscopes (ESEMs). After the electron beam hits the surface measurements are made of secondary electrons or X-rays that may be given off by the surface after the electron interaction. These emitted energy sources are then entered into a computer program which creates the image of the surface. The operation diagram of an SEM is shown in Figure 3.8.

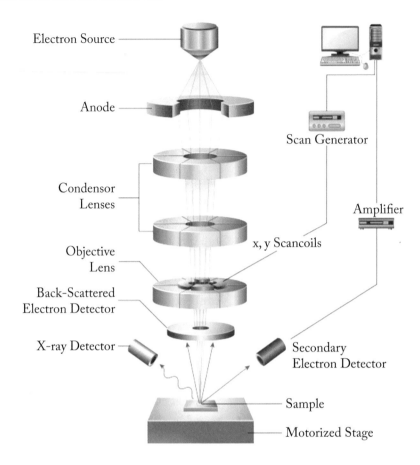

Figure 3.8: Operational diagram of an SEM.

Other tools used to study the nanoscale include X-ray diffraction equipment and Raman spectrometers. Both of these tools can be used to evaluate the crystalline and molecular structure of materials.

A much larger and complicated tool which is finding more use in nanoscience is the nuclear magnetic resonance equipment shown in Figure 3.9. This equipment is used to study the nuclei of materials and associated properties using two interacting magnetic fields. This equipment is usually applied to specific measurements and property definition after significant preliminary work has been performed using the other tools of nanotechnology.

Figure 3.9: A nuclear magnetic resonance machine.

Finally, a nanotechnology investigation tool which is slightly different from the previously discussed tools of nanotechnology is the nanomechanical indenting tool. This tool has predominantly been manufactured by the Hysitron Corp. (now a subsidiary of Bruker) and is shown in Figure 3.10. This tool is unique in the sense that it provides quantitative measurements of the physical properties of a material. The indenter operates using a cantilever beam similar to that which is used by an atomic force microscope. The tip is very carefully geometrically characterized. The tip is used to indent the material under controlled pressure, environment, and timing. By this method the mechanical properties of materials can be very accurately measured. The result of the indentation is a graph representing the force vs. the displacement. This resulting graph is shown in Figure 3.11.

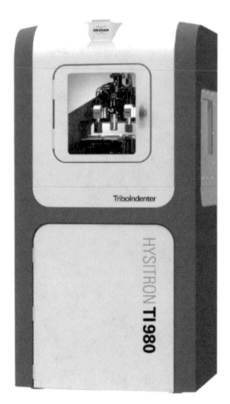

Figure 3.10: Hysitron/Bruker nanomechanical indenter.

Figure 3.11: Results of a nanomechanical indentation in fused quartz.

Analysis of various sections of the resulting graph can provide information about the specific mechanical properties of the material under testing such as reduced modulus and hardness. Other information such as stiffness, delamination, film thickness, and fracture toughness can also be obtained. The right-hand portion of Figure 3.11 is an image of the resulting indentation in fused quartz used as the sample material.

As interest in nanotechnology grows, the tools used to observe, measure, and study at the nanoscale will continue to evolve. One of the most significant evolutions is the creation of nanoscale evaluation tools that are smaller and less expensive than the research-grade tools. These newly developed tools are especially compatible with undergraduate, community college, and high school courses. These tools will allow an extended and younger audience to learn and apply nanoscience to multiple disciplines.

There are additional tools of nanotechnology which, rather than allowing for the investigation of individual atoms, allow for the differentiation and determination of the specific atoms that may compose a compound. These tools are the mass spectrometer and the Raman spectrometer.

This form of analysis and testing has been available for a long period of time, with the first mass spectroscopy measurement being performed in 1912. Mass spectrometry involves molecules in a vacuum and in a gaseous form moving through a column of electrons. In this interaction the molecules are "broken" apart into sub-elements consisting of molecular ions and fragment ions or sub-molecules. Because of the interaction with the electrons the resulting sub-molecules will carry a charge. These charged molecular portions are then passed through a magnetic field. The magnetic field will cause the path of the molecular ions or fragment ions to be modified based on the charge and predominately the mass of the particle. Because of the variation in mass, the resulting sub-molecules will interact with a detector screen at different locations, allowing them to be measured as separate entities. As the different sub-molecules hit the screen and using a computer program, they will be "measured" in terms of the mass to charge ratio (m/z) and the abundance of the particle having that same mass to charge ratio. The results are then plotted with the m/z value plotted on the horizontal axis and the relative abundance, when compared to the most abundant sub-element, on the vertical or y-axis. In this manner the composition of complex molecules can be determined.

Raman spectroscopy is named after the scientist Chandrasekhara Venkata Raman, who, in 1921, wanted to find the explanation for the blue color in glaciers and started experimenting with light interacting with different forms of water. Raman spectroscopy, similar to mass spectroscopy, also deals with materials in their molecular form. The Raman system uses light, usually higher-energy laser emissions, to interact with the molecules. The samples can be in liquid or solid form and are usually placed on a glass slide. Raman spectroscopy does not require a vacuum and one of the benefits is that the sample material is not destroyed so it can be used for additional experiments. This method can be used to define the chemical structure, phase, crystallinity, and molecular interactions of a material. The high-energy laser light shone on the sample results in the majority of the resulting light being of the same wavelength as the impending light, known as Rayleigh scattering. A small portion of the light is scattered and emerges from the interaction with a different wavelength. This is known as Raman scattering. By observing and measuring the resulting, emitted Raman scattering light, that the properties can be determined after the interaction. The result is a series of peaks with the wavelength plotted on the x-axis and the intensity on the vertical, y-axis. Each peak corresponds to a different molecular bond. The complete resulting Raman spectrum is very distinct for individual compounds and allows for a quick determination of the chemical composition. For example, the spectrum for ethanol, CH_3CH_2OH, is different from the spectrum created by methanol, CH_3OH. Libraries of spectrums have been created allowing for a quick assessment of a given sample.

Finally, X-ray diffraction is another experimental tool often found in nanotechnology labs. As the name implies, this tool uses X-rays to illuminate and study a crystalline structure and the resulting diffracted x rays are viewed on a screen to determine the atomic and/or crystallographic structure of the material. X-rays have wavelengths in the range of 0.1 nm (the diameter of a hydrogen atom) up to 10 nm. The energy range is from 100–200 eV. With the short wavelengths, X-rays can interact with materials at the atomic level. The X-rays impinge upon a material and will penetrate and interact with multiple atomic layers. They will then be diffracted by the atoms and emerge from the sample with the paths of the waves modified and of different lengths because of the interaction with the different layers of atoms. These paths of the emerging X-rays will interact with each other resulting in either constructive or destructive interference. On the detector screen, regions of brightness will appear. By measuring the separation of these regions, information about the atomic structure, i.e., distance between layers and separation of atoms, can be determined. This information provides a detailed assessment of the crystalline structure of the material. X-ray diffraction is performed in a vacuum and requires additional worker shielding and protection.

CHAPTER   4

Society and Nanotechnology

4.1   INTRODUCTION

In this chapter, the secondary impacts of nanotechnology and society in general are covered. There are several aspects of modern technology, including the Internet, that have impacted how society and the world have responded to the new, emerging technology that is nanotechnology.

4.2   THE GLOBAL NATURE OF NANOTECHNOLOGY

There are several aspects of nanotechnology which truly make it different from previous technology revolutions. First, nanotechnology is global. Almost every country has some aspect of research that involves nanotechnology or takes advantage of the properties and attributes of the nanoscale. Second, society or the public, in general, has become aware of nanotechnology much earlier in its developmental cycle than the time when the public is traditionally aware of emerging or new technologies. Another aspect of nanotechnology which is extremely significant and will be discussed in detail is the fact that one development at the nanoscale can have implications in multiple disciplines. This includes technical and non-technical disciplines as well as having multi-market applicability. Finally, the multidisciplinary aspect of nanotechnology which was covered in Chapter 1 will impact the development, applications, and societal implications as well as awareness and implementation. The multidisciplinary aspect also poses a substantial issue for both businesses that are trying to develop products that involve nanotechnology as well as the educational environment.

Nanotechnology is indeed global. Figure 4.1 shows the increase in nanotechnology funding and development for different countries over a 15-year period as it relates to the U.S. investment.

As is shown in the figure, the U.S. is not the only country investing substantial amounts in nanotechnology. Countries such as China and Japan have substantial investment in the nanotechnology arena and India, although not shown in this figure, is quickly escalating nanotechnology investment. Multiple countries are also integrating nanotechnology into their educational library. Industries are also investing in nanotechnology in multiple countries throughout the world, creating new partnerships and companies.

The aspect of intellectual property protection and patent law will be discussed later in this chapter.

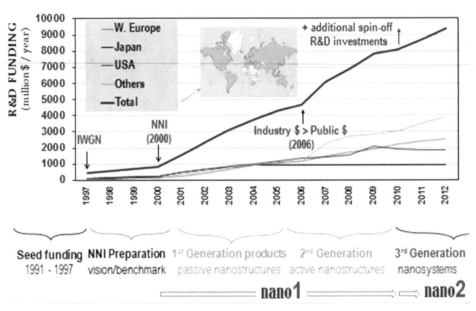

Figure 4.1: Research and development nanotechnology spending by country.

4.3   SOCIETAL AWARENESS OF NANOTECHNOLOGY

Figure 4.2 shows traditional "S"-curves. This type of curve, an "S"-curve, is most often referred to by economists.

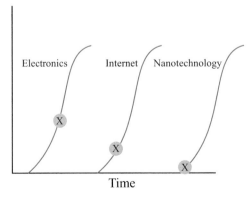

Figure 4.2: Representative "S"-curves.

An S-curve is created when time is plotted on the horizontal axis and cost or number of products or market size is plotted on the vertical axis; it is a measure of the strength of the product or market or the acceptance by the public of that given product. At the beginning of product initiation or product announcement the market is at the bottom of that S-curve. Over time, as public awareness increases and more products are sold, the product begins to move up the S-curve. As the product becomes more widely known and gains popularity, movement occurs into the vertical part of that S-curve. In this portion of the curve lots of products or units are being sold and the market is very strong. The product experiences wide publicity in the U.S., for example, and people in multiple countries are aware of the product. This is the period of time when a product for the company is going to be most profitable. As time progresses, the top portion of that S-curve is reached and a company will begin to sell fewer products and the profitability is not as great. At the top portion of the S-curve, in many cases, there's another competitor or a new variation of the product or perhaps a brand-new product. It is usually the desire of a company to have overlapping S-curves so that as one product is becoming obsolete or outdated another product is in development or available to take its place and take over that vertical segment of the S-curve.

Figure 4.2 shows a series of three S-curves; one for electronics, one for the Internet, and one for nanotechnology. The "X"s in the figure represent an estimate of the time when the public became aware of that technology. For the electronics market, the transistor was researched, developed, and applied long before the public became aware of, for example, transistor radios. The transistor existed while most homes still had vacuum-tube TVs and radios. It existed long before the public became aware of electronics and what those electronics could do. The same is somewhat true for the Internet. The Internet was developed under Department of Defense funding and in many cases the Internet was in use by government personnel prior to the public being aware of the Internet.

However, in the case of nanotechnology, most likely because of the first two S-curves, electronics and the Internet, the public is very much aware of nanotechnology. The public may not be aware of the exact definition or the scientific discipline of nanotechnology but in some cases, because of popular science fiction books and articles that are published in newspapers, the public is aware of nanotechnology. Most likely the public does not have the technical understanding of nanotechnology but, again, the public is aware of the phrase and the potential, both positive and negative, of nanotechnology.

In the early days of nanotechnology, low on the S-curve, the majority of the work was in scientific fields and was research related. During this early phase the tools were very expensive and the concepts were challenging. Therefore, early work in the nanotechnology field was performed by large companies such as IBM, Hewlett Packard, and Motorola or by large, well-funded universities. The effort and emphasis were in terms of understanding the world at the nanoscale and developing and improving the tools to observe the world at that scale. The majority of the activities that involve nanoscale early on or low on the S-curve did indeed involve research aspects rather than specific application goals. Some of the properties under investigation are shown in Table 4.1. Researchers were attempting to understand the nuances and changes in physical, electrical, and biological properties as the size changed.

4.4   THE SLOW RISE OF NANOSCALE APPLICATIONS: A COMBINATION OF TECHNOLOGY, REGULATION, AND ENVIRONMENT CONSIDERATION

There was and has been minimal application to products and procedures and, in general, nanotechnology has not moved too far up the S-curve in terms of return on investment, financial viability, financial benefits, or product development and broad application to the public. Many people thought that the nanotechnology "revolution" would be similar to the dot-com boom and expected rapid growth and major financial benefits. This turned out not to be the case.

Scientists were evaluating aspects shown in Table 4.1 and attempting to understand nuances of physical properties and interactions as shown for long periods of time. Often, as previously mentioned, it has been found that subtle variations in purity or size and shape can have a significant impact on the physical properties.

There are many factors that are contributing to the slower ascension of nanotechnology on the S-curve. A predominant contributor is the fact that the equipment required to investigate at the nanoscale is expensive, as discussed in Chapter 2, and in many cases non-trivial to operate. A second factor is that the number of skilled equipment operators is low. Many of the personnel with knowledge of the equipment operation hold Ph.D.s and therefore command a high salary, making it difficult for small or entrepreneurial companies to move into the nanotechnology arena. Finally, the interpretation, assessment, and evaluation of results from the nanoscale test equipment may vary dependent upon multiple factors such as test equipment, environment, approach, sample purity, preparation etc.

There are four other factors that have contributed to the slow rise in the impact of nanotechnology on companies, in products, and with the public. First, it has been found that properties of materials are dependent on the size of the material. Table 4.1 presents a list of the properties being studied that have been found to change as the size of the materials is reduced. For example,

Table 4.1: Some properties under investigation at the nanoscale.
Property
Strength
Ductility
Melting point
Electrical strength
Magnetic attributes
Hydrophobicity
Friction and adhesion
Toxicity

a piece of silver, such as a coin, is not magnetic. However, if a nanoparticle is created with a composition of only 13 atoms of silver, that nanoparticle has magnetic properties. Property changes at the nanoscale can result in new applications and products. These changes can also pose challenges to manufacturing, measurement, and produce unintended results. The unintended results and the property changes also present regulatory challenges. A question often arises as to whether the patent or regulatory requirements defined for a large quantity of material is applicable to a nanoscale quantity of the material.

Hence, the regulatory aspects are the second factor that is contributing to the slow introduction of nanotechnology into products and use. It is also important to note that the intended use of the nanomaterial will impact regulatory aspects. For example, if a manufactured nanoparticle has an intended application as a plant fertilizer the regulations applied for that material may be different than the regulations if the same material has a proposed use in drug delivery.

Third, it has proven, in most cases, to be expensive to manufacture nanoscale materials, especially when the desire is for the resulting product to be similar or uniform within a manufactured "batch." This is especially true in the area of medical applications. Experimentally, it has been shown that gold nanoparticles can be coated with various biological and pharmaceutical entities allowing these engineered nanomaterials to be used to treat various diseases such as cancer. This application requires that the starting material, the gold nanoparticles, be of a specific purity and size. Creating a uniform population of gold nanoparticles is a significant challenge, in addition to the process of measuring and qualifying the resulting coated gold nanomaterial product.

Finally, the fourth factor involves toxicity and/or end of life. The newness of nanotechnology and the creation of a multitude of engineered nanomaterials (ENMs) contributes to the lack of knowledge regarding potential toxicity and impact to the environment. One of the challenges in this area is how to account for the effect of combined accumulation of nanomaterials. For example, one factory may be producing golf clubs with CNTs in the shaft. Although most of the CNTs may indeed end up in the formed golf club shaft, there will probably be a small percentage of the CNTs that end up in wastewater or waste material. This particular factory may be contributing to a small, regulatory agency-allowed amount of nanomaterial into the environment (water, landfill, or air). There may be companion factories in the same region also contributing nanoscale material to the environment. In this scenario, each factory is meeting the constraints of regulatory levels, yet the potential for the combination of all of these contributions at a central location, resulting in a mass accumulation that could pose a hazard exists. This potential accumulation of engineered nanomaterials has not been addressed well. Granted, the same regulatory issue exists for any potentially dangerous or toxic material when multiple entities are each contributing only a small amount of hazardous material, but the potential exists for all of that material to be combined at a central location. Another example involves the medical application of nanoparticles. In the example above, gold nanoparticles were used as a core material that could be coated with a protein or antigen to detect malignant cells and could also be coated with a pharmaceutical that could kill the malignant cell. The consideration here is to define what happens to the nanoparticle after it has performed the desired function, i.e., find and destroy the malignant cell. The fate of the nanoparticle could be to remain in the body or be eliminated in urine, fecal material, or breath. If the nanoparticles end up in a waste treatment facility, the subject nanomaterials could be combined with other materials. This scenario is similar to what is currently happening with pharmaceutical drugs. Not all of the medical material is used by the body and ends up in treatment facilities where is could find ways into public water or food supplies. The question involves not only the initial core material of gold but could also involve the consideration of any residual target or pharmaceutical entities that may have been used to coat the gold core.

Although these issues of hazards or toxicity have been brought to the forefront because of the modern era of nanotechnology, nanoscale entities have existed in our environment for centuries. Smog, sprayed pesticides and fertilizers, and diesel engines have all contributed nanoparticles to the air, water and soil. Also to be considered is the safety of the workers involved with the creation, measurement, and packaging of produced small-scale materials. These materials may not be pure nanoscale but are small enough to be inhaled or penetrate through the skin, very similar to the small particles that exist naturally. The workers dealing with small materials often work in controlled environments with a defined air flow, protective clothing, and often under fume hoods. The safety of these workers is under consideration by several organizations such as the National Institutes of Health and the Environmental Protection Agency.

The reasons discussed above, as well as the fact that it is difficult to imagine entities at the nanoscale, have resulted in a moderately slow introduction of nanomaterials and products based on the benefits of nanoscale materials.

4.5   CARBON NANOTUBES: THE ICON OF NANOTECHNOLOGY

The fact that one nanotechnology development or discovery can have applications in multiple markets poses a challenge for educators, businesses, and the public. The interpretation of a nanoscale discovery can quickly become complex and complicate the process of understanding the nano-entity and what positive or negative potential exists. As an example, consider the CNT, shown in Figure 4.3 below.

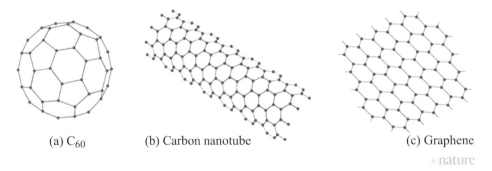

(a) C_{60} (b) Carbon nanotube   (c) Graphene

Figure 4.3: Carbon nanoscale structures.

A CNT, shown in the center diagram, is a hollow tube composed of carbon atoms. The CNTs may be of varying diameters. In Figure 4.3, the carbon atoms are represented by the small spheres. The chemical (covalent) bonds between the carbon atoms are represented as the lines joining the atoms and cannot physically be observed. A carbon atom has four electrons in the outermost orbital. This implies that a carbon atom could potentially bond with four other atoms. In carbon structures such as C60, CNTs, and graphene (as well as benzene), the carbon atoms will bond with only 3 adjacent atoms, because of the substructure of the orbitals. Remember this is a representation of the structure and the chemical bonds that are represented by the lines are not truly visible.

Figure 4.3 shows a single-walled carbon nanotube (SWNT) with one layer of carbon atoms. In many cases, the "rolls" of carbon atoms will be inside of each other resulting in multi-walled CNTs that consist of anywhere from 2–6 different walls of carbon atoms. Multi-walled nanotubes are easier to manufacture than the single-walled CNTs. The CNTs can be twisted, like a roll of chicken wire, and with this twisting the·CNTs can exhibit different electrical properties. This twisting results in the hexagons of carbon atoms being arranged differently and results in different electrical properties.

CNTs also can occur in different diameters with the smallest CNTs being around 3–4 nm in diameter. One of the challenges of the manufacture of CNTs is creating CNTs that have a substantial length. Currently ,CNTs can be manufactured with lengths of several millimeters. The assembly and manufacturer of CNTs has been a nontrivial exercise and there have been several companies who have initiated the business and then gone out of business because of the challenges associated with the manufacture of CNTs. Especially desired for many applications are CNTs that have a level of purity and a specific orientation. Again, that orientation provides the physical and electrical properties of any given CNT. However, CNTs with different orientations, lengths, and numbers of layers or rolls are applicable for many composite material uses where the main benefit of the CNTs is to increase material strength while reducing weight. Yet, the manufacture of CNTs has been a substantial challenge.

4.6   AN EXAMPLE OF THE MULTI-MARKET ASPECT OF NANOTECHNOLOGY

The CNT discussed in the last section is a perfect entity to use as an example of the multi-application aspects of a nanoscale discovery. The Buckyball (C_{60}) was discovered in 1985 as a result of Sir Harry Kroto wanting to understand the interstellar readings he was observing and relate those readings to a particular molecular structure. Rick Smalley, Robert Curl, and others at Rice University had a capability to create various atomic structures using their ablation experimental apparatus. It was through a series of experiments that the C60 molecule was created which led to the CNT structure and eventually to the graphene structure discovery. Figure 4.4 is a representation of the multi-applicability of CNTs.

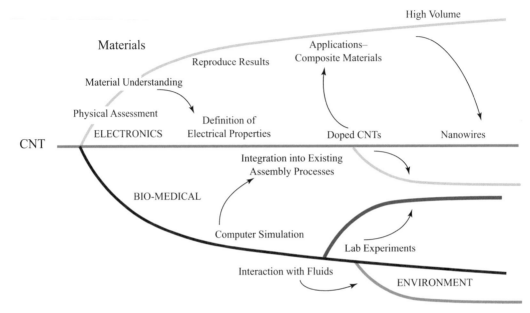

Figure 4.4: Multiple applications of a single nanoscale entity.

Starting at the left-hand side of Figure 4.4, we have the CNT represented. Following the top line is a representation of the earliest measurement and understanding of CNTs and dealt with the material or physical properties of a CNT. The CNTs that were initially created looked like soot on the bottom of the vacuum chamber. Using the experimental chamber at Rice University, the Buckyball was discovered and also created were the various carbon "tube" structures.

Upon the realization that the soot contained these tube-like structures, the effort by research scientists, predominately Rick Smalley and his group at Rice University, was to try and understand the physical properties of the material observed at the bottom of the vacuum chamber. Within the materials science category, shown in the top portion of Figure 4.4, was the effort to study and understand the physical properties of the material. This understanding would support the initial definition of electrical, physical, and thermal properties of this nanoscale material. The effort continued by moving further up that line and included the reproduction of those results as well as continued measurement and evaluation. When evaluated under certain microscopes, like the atomic force microscope or the SEM, variations were observed in the produced material such as tubes within tubes which became known as the previously mentioned multi-walled carbon nanotubes (MWCNTs).

Over time, the efforts transitioned from research lab creation of the CNTs to factory or higher-volume production of the material. Initially, the high volume of CNTs created were independent of the orientation or chirality, the size of the CNTs or purity of the CNTs. The initial effort was to determine if the process of creation could be replicated and if it could be replicated in high volume. Over time the effort has focused on creating CNTs in high volume that have specific properties whether those be physical, electrical, or thermal properties. Purity of the produced material was also a consideration. Continuing to follow the top line, the effort to take advantage of the material properties of these CNTs was initiated. Today CNTs and their material properties such as very high strength as compared to their weight is being applied to tennis rackets, grommets, and building materials, especially concrete. It has been found that when CNTs in very small percentages are integrated into concrete the result is concrete that has substantial or equivalent strength for about ½ to ⅓ of the weight, thickness, and volume of concrete needed. This has a tangential impact because the creation of cement, which is a critical component of concrete contributes significantly to air pollution. The use of CNTs to strengthen a thinner amount of concrete will reduce the amount of cement required and as result reduce air pollution.

The next application of CNTs follows the middle horizontal line labeled electronics. Not long after the discovery of CNTs and the measurement of their material properties it was determined that CNTs had characteristic electrical properties. It was found that some CNTs served as electrical insulators, others as conductors, and some as semiconducting materials. After a period of investigation, it was determined that these electrical properties were dependent upon the chirality or the twisting of the CNT. Using the example of a roll of chicken wire representing our CNT, twisting that chicken wire one way you may end up with an electrically conductive material. Twisting another way may result with an insulating material. Therefore, different electrical properties are determined by the chirality or the twisting of the CNTs. The precise understanding and mechanism for those electrical properties was undefined initially and is still under investigation at the atomic level. However, manufacturers have learned how to control the manufacturing process such that the twisting of the CNT is defined. Hence, specific electrical properties of the resulting produced CNTs can be defined.

CNTs have the potential to make ideal transistor material. For example, transistors using CNTs could be made extremely small, therefore they would have the potential to be very fast, switching transistors. The power requirement may also be significantly reduced. However, the challenge of applying CNTs to electronics lies in the assembly of transistors from the CNTs. Transistors are extremely well-ordered arrays of different materials and controlling the arrangement of the CNTs has proven to be a challenge. (Remember CNTs are extremely small and require specialized microscopes to observe and measure them. It is not like laying down a row of pencils!) Another challenge facing the electronics industry is obtaining CNTs that are identical, either in terms of length or diameter or chirality. In order to create a transistor device, it is required to have multiple CNTs with identical electrical and physical characteristics.

These issues created significant challenges for electronics and electrical engineering research-ers. One solution, which is implemented at the device level, is to use various fault tolerant architec-tures that can be adapted to account for erroneously placed or incompatible CNTs in the transistor array. In many cases redundancy of individual device sections or segments which perform a specific function is used. Software is implemented such that if the operation fails in the first segment the function would be switched over to the secondary or back-up circuit segment. Because of the po-tential for very small transistors, resulting in small circuit elements that perform a specific function, multiple backup, or redundant circuits could be implemented. Another approach that was used, as shown in Figure 4.4, is to "dope" the CNTs with other elements or atoms. In this case, elements, other than carbon, were added to the manufacturing process. The CNTs now contained elements other than carbon. Through chemical or electrostatic interactions, the assembly, location, and plac-ing of the CNTs could be controlled. To a certain degree the electrical characteristics could also be controlled using the "doped" CNTs.

Finally, some researchers investigated the CNTs as forms or molds to be used to create very small wires that could be used in electrical devices. CNTs have an inner diameter of 2 to 3 to 6 nm, therefore providing a good mold to create nanowires. Nanowires are difficult to manufacture by traditional wire fabrication methods when very small diameter wires are required.

The bottom line in the diagram of Figure 4.4 shows the application of CNTs to biomedical or medical fields. Carbon is biologically compatible and having that material in the form of very small, narrow hollow tubes led to an interest in using CNTs as vein or capillary replacements. Be-cause of the small size and biocompatibility, it was considered that CNTs might be able to be used to create artificial kidneys or livers.

However, what was discovered was that the flow of liquid, whether water, blood, or serum in CNTs with very small diameters did not respond the way that fluid flows in larger systems; see Figure 4.5. Meniscus shapes and flow rates did not correspond to traditional equations.

As can be seen in the figure, the shape of the water/vapor interface is anything but what is the traditional slightly curved interface. The curve that is traditionally seen is only because the adhesive forces between the container and the water are usually stronger than the cohesive forces between the water molecules in the liquid. And, as seen in some cases in the figure a meniscus may never form. The CNTs in Figure 4.5, (a)–(e), are fairly large diameter tubes of approximately 200 nm, based on the scale bars. Figure 4.5 (f) is a smaller diameter tube of approximately 50 nm. A water molecule has a diameter of 0.275 nm, a small molecule, so quite a few water molecules could span the diameter of the CNT.

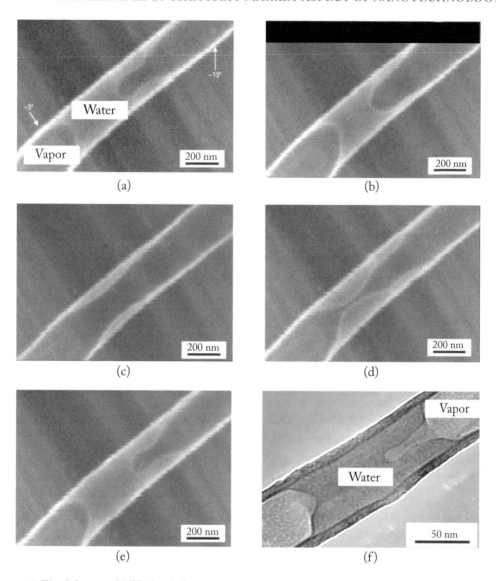

Figure 4.5: Fluid flow in CNTs for different temperatues and pressures.

The medical field continues to evaluate various applications for biologically compatible CNTs and other compatible materials such as DNA. Both CNTs and DNA are being investigated as scaffolding materials. The scaffolding would serve as "bridges" or supporting structures to speed the growth of cells and tissue. For example, it takes a considerable amount of time for bones to heal, predominately because the bone cells are long structures and it takes substantial time for the bone cells to reconnect and form a strong structure. Use of CNTs, which can be narrow and long, can serve as a supporting structure for the bone cells, thereby speeding up the regrowth process.

CNTs are also being evaluated as potential pathways to regrow neurons. Caution needs to be applied because CNTs, unless coated, can carry an electrical charge, because the uncontained extra fourth electron not covalently bonded to another carbon atom in the tube will result in a non-uniform charge distribution. The electrical charged entities can damage biological materials such as cells. Therefore, CNTs that are proposed for use in biological systems need to be coated or have the charge neutralized in some manner.

4.7   LEGAL CONSIDERATIONS

Entering the era of nanotechnology has resulted in new scientific understanding, new materials and processes, as well as new products. Research scientists may want to protect their discoveries and companies want to protect any new created materials, processes, or products. Within the U.S. there are several approaches toward protection of your knowledge or product: (1) company secret or proprietary, (2) trademark, (3) copyright, and (4) patents.

First, a company secret or proprietary designation is often used for a process developed within a company that results in a particular material or property within a material that is intended to be kept secret. That is, no one outside of the company, or a small group within the company, is aware of the discovery or process and no one is allowed to share the secret outside of the company or a designated group of people within the company. An example of this approach is the Donaldson Company. One of the major products of Donaldson is large air filters. These filters are used in tanks and ships in the military, large buildings, and medical facilities. Many years ago, a technician at Donaldson developed a process to create very, very thin fibers—nanoscale fibers. Using these fibers in air filters would allow for much smaller particulates to be removed from the air and provided an advantage over their competitors and a benefit to their customers. The process to create these fibers was kept a company secret for over 15 years. The products that used fibers created by this process were sold and the names trademarked (protected) but the actual process to create the fibers was not disclosed. When a new discovery is made, the initial approach is usually to designate that approach, material or process as a company secret.

Second, a trademark is legal protection for a name, symbol, phrase, word, or design. Having trademark protection prevents anyone else or another company from replicating, copying, or using an entity that is moderately close to the one you have trademarked. Filing for a trademark is done by filling out an application form with the U.S. Patent and Trademark Office (USPTO).

Third, a copyright usually applies to written work. This could be papers published in journals, magazines, or books. Copyright is also applied to presented content such as may be found in PowerPoint files. In some cases the copyright may belong to a specific person and in often for technical journal publications the copyright belongs to the published entity such as the American Chemical Society (ACS) or the American Association for the Advancement of Science (AAAS).

The copyright protection pertains to both the written text, any equations and graphics, images, or drawings in the publication. Use of copyright material usually requires approval from the copyright owner, except when the material is going to be used for educational purposes and is not intended to be sold.

Finally, a patent is designed to protect materials, a process, or a new product and provides the property rights to the inventor. The application form for a patent is non-trivial and the approval process within the USPTO can take up to three years. Once a patent is filed it becomes available to the public through a searchable data base. For this reason, many companies may decide to keep a new development as a company secret for a period of time before filing a patent. Also, patents may be filed either as an offensive move or a defensive move. From an offensive standpoint, the patent outlines/defines the process or product that is intended to be produced and sold. Defensive patents may be filed to protect against competitors who may develop slight modifications and produce a product similar to the patented design. An example of such an approach centers around the development of a new version of a spacecraft control computer. For many years these electronic systems had been designed using custom circuits for control and processing. The engineering team proposed a new design that was based on newly developed commercial devices rather than the custom designed circuits. The new proposed approach had many benefits. After a period of time of testing, assessing, and validating the new approach to a certain degree, two patents were filed that represented the new approach and design. However, during the process of developing and designing the new system, several alternate approaches were proposed which could have the same final result but for other reasons were rejected. Three patents were filed for these alternate designs as a defensive posture against anyone who may want to use these secondary approaches.

There are multiple ways to protect intellectual property as discussed. There are significant challenges associated with protecting nanoscience and nanotechnology developments starting with the question regarding the newness and the ability to protect a molecule.

CHAPTER 5

Investigating the Relationship: Is Nanotechnology a "Basic" Science or Are the Traditional Sciences Nanoscience?

5.1 INTRODUCTION

Many traditional scientists and educators will say that they have been doing "nanotechnology" their entire career. In the purist sense this is true—all of the traditional sciences; physics, chemistry, and biology have dealt with the world at the atomic or molecular level to one degree or another. However, the recent advent of "nanotechnology" or "nanoscience" has placed a new emphasis on viewing, studying, and acknowledging the world and all of the materials and interactions at that level. Note that often material science is included as one of the basic sciences as is geology or earth science.

Use of the tools developed in the recent history of nanotechnology has allowed for clearer understanding and sometime verification of concepts covered in these basic science disciplines. There is a shift in thinking about how everything from metals to polymers to proteins and bacteria are viewed. Shining a light on the nanoscale as those aspects and concepts relate to how the world is traditionally viewed is the focus of this chapter. There are multiple sections in this chapter, including finding nanoscale concepts in earth science, physics, chemistry, and biology; it all boils down to "opposites attract and likes repel."

5.2 FINDING NANOSCALE CONCEPTS IN THE BASIC SCIENCES

First, a reminder from the first page (Figure 5.1) that everything is made of atoms and atoms are composed of electrons, neutrons, and protons. And the fundamental rule of the interaction of atoms is that opposites attract (electrons and protons) and likes repel (electron to electron and proton to proton). This fundamental concept is critical when evaluating the physical, electrical, and biological properties of any material.

For example, consider the carbon atom, which is the atom diagramed in Figure 5.1. Carbon atoms are fundamental to biological systems and are the building blocks of many materials. A classic example of the importance of the atomic arrangement of atoms is presented in Figure 5.2.

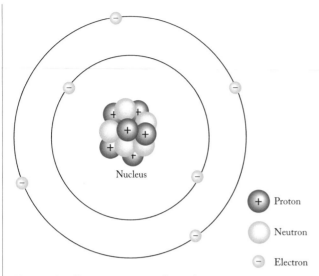

Figure 5.1: Representation of a carbon atom.

Carbon atoms arranged one way result in a mass of charcoal. Subject those same carbon atoms, in various forms, to high temperatures and pressure, which stress the bonds between the carbon atoms and modify the placement of those atoms, and the result is a diamond. This insight into the structure of different arrangements of carbon atoms validates and explains aspects of physics, chemistry, and material science. As stated previously, a stable balance of energy is obtained when eight electrons occupy each atomic energy level which is designated by the outer ring in Figure 5.1. The carbon atom has four electrons in that outer most ring. Hence, the carbon atom is an ideal candidate to merge with four other atoms, if each atom contributes one electron leading to its prominent position in chemical interactions and multiple molecules. Carbon, being biologically compatible, plays a significant role in many proteins, fats, and even DNA.

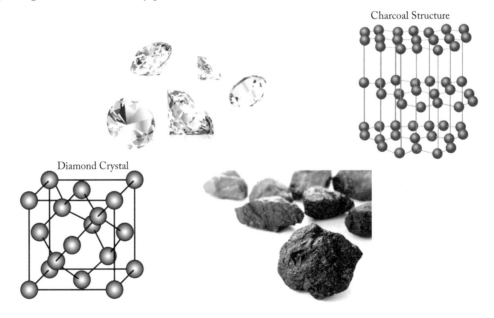

Figure 5.2: Carbon atoms with different structures resulting in either diamond or charcoal.

Often adjacent atoms contribute more than one electron, which is the case in CO_2, carbon dioxide. In this case each oxygen atom, with six electrons in its outermost ring will share two electrons with carbon, similar to the ionic bond which occurs in salt ($NaCl$). This sharing of two electrons results is a "full" ring with eight electrons for each of the oxygen atoms and eight electrons (four shared) in the outer ring of the carbon atom.

The above discussion using a carbon atom as an example of how atomic structure or the nanoscale impacts and helps support the understanding of the electrical, physical, and biological properties of materials. This critical relationship and dependency are represented in Figure 5.3.

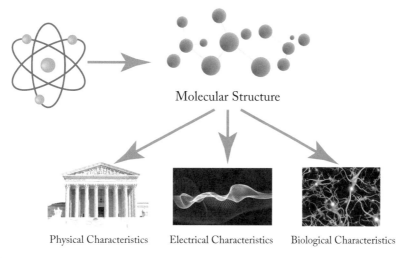

Molecular Structure

Physical Characteristics   Electrical Characteristics   Biological Characteristics

Figure 5.3: Atoms to molecules to properties.

5.3   EARTH SCIENCE

The subject of earth science can be as narrow or broad as one would like. From a geology standpoint, the tools of nanotechnology have allowed researchers to determine the atomic level structure of sediments and rocks. This understanding has led to a clearer understanding of planetary history and plate tectonics as well as earthquake and volcanic activity. This understanding of our home planet will help increase the understanding of other planets within and beyond our solar system as well as understanding the sun and other stars. This fundamental understanding may someday lead to exploration of planets and mineral and mining operations, as well as providing new insights into energy production.

Earth science also encompasses weather and atmospheric phenomena. Clouds, rain, snow, and rainbows are all the result of interactions at the atomic or nanoscale level. Raindrops form because of the hydrogen bonds that exist between water molecules but also because dust and other particulates in the air are often charged with either a positive or negative charge. These charged particulates will then interact and attract water molecules to form raindrops. Lightning is clearly a powerful example of opposites attracting where there is a large negative charge, for example in the clouds, and a large positive charge at the surface of the earth. Lightning most often occurs within the clouds where regions of positive and negative charge have built up and the interaction or attraction between those two charged regions results in lightning.

Even the subatomic particles that constantly bombard the surface of the earth are a result of a nanoscale interaction between ionized (charged) atoms from either the sun or outside of the solar system, sometimes known as cosmic rays, and the earth's atmosphere. Atoms of oxygen, nitrogen, and other gases in the earth's atmosphere will interact with the ionized particles impinging upon the upper atmosphere. This shower of particle interactions is shown in Figure 5.4.

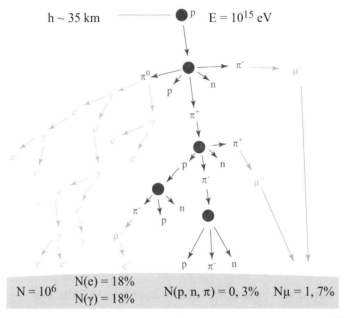

Figure 5.4: Example of secondary particles generated by a proton entering the earth's atmosphere.

This cascading of molecular or atomic level interactions will result in particles landing on or passing through the earths surface, such as neutrinos.

Also, particles with origins outside of the earths atmosphere are responsible for the northern and southern lights. As these atoms near the earth they will interact with the earths magnetic field. The interaction with the magnetic field will cause electrons in the atoms to change orbitals or energy levels. This change in orbitals can result in the release of energy which is often seen as the various colors in the lights at the earth's magnetic poles. Different colors observed in these lights are due to electrons changing orbitals in different atoms. Each atom has a different electron structure and therefore electrons will lose different amounts of energy, resulting in the different colors observed.

5.4   BIOLOGY

Nanotechnology, and the fundamental interaction of opposites attracting and likes repelling provides the foundation of biological interactions.

The charged entities that form the nucleic acids in DNA, the structure of proteins and enzymes, and the structure of antibodies, cells, and other entities are all dependent on this interaction. Several of these structures, as well as a water molecule, are shown in Figure 5.5. The predominant atoms are carbon, oxygen, hydrogen, and nitrogen. As shown in the figure the atoms are very close together because of the interaction of the electrons and protons in the atoms. The reason that proteins fold the way that they do and drugs interact with various portions of the proteins is because of the way the charged regions around sections of both the drug and the protein interact.

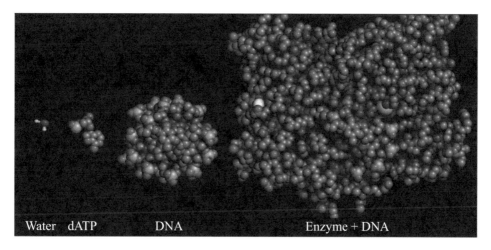

Figure 5.5: Computer simulations of various biological molecules.

There is substantial interest in using nanoparticles as modes to detect diseased cells and deliver a drug. This application serves as an example of the impact of nanotechnology on all areas of basic science.

Consider a gold nanoparticle which is to be coated with a detection protein or enzyme (shown in purple in Figure 5.6) to find and attach to a specific diseased cell, perhaps a cancer cell. The nanoparticle is also to be coated with a drug or molecule (shown in green in the figure) that is toxic to the cancer cell.

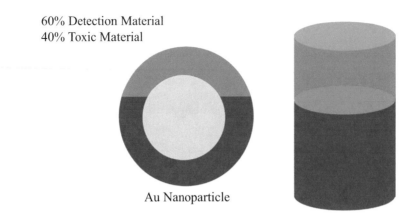

60% Detection Material
40% Toxic Material

Au Nanoparticle

Figure 5.6: Nanoparticle with detection and drug material coating and a beaker of a solution of coated nanoparticles.

Also assume that the desired ratio of detection coating to drug coating is 60–40. The coated nanoparticle is shown in the left side of Figure 5.6. The right-side portion of Figure 5.6 shows a beaker of the nanoparticles in a solution. This would most likely be the vehicle by which the coated nanoparticles would be introduced into the patient, that is, in a solution.

One of the critical aspects of the development of this dual-purpose nanoparticle is the assessment or measurement of the nanoparticle at the end of the assembly process. It is necessary to ensure that the starting gold core of the nanoparticle is of the correct diameter and also that the coating, of either type, is also of the correct thickness. The human body is designed such that entities that are very small are quickly eliminated from the body and entities that are larger are viewed as invaders and attacked by macrophages or antibodies. This is how the body would attack an invader such as a splinter. Because of these size constraints imposed by the human body it is necessary to "manufacture" these coated nanoparticles within a specific size range.

Not only is the size of the finished coated nanoparticle of importance but also the desired ratio of the detecting element and the drug element of the coating needs to be measured and ensured. The nanoparticle approach would be inefficient if the nanoparticles were nearly 100% of either the detecting element or the drug element. When the produced nanoparticles are placed in a solution in a beaker the percentage of detection material and percentage of drug material can be measured at the macro level. However, ensuring that each nanoparticle meets the specified requirements is a much more difficult task. The process to measure each nanoparticle would most likely require an SEM or a transmission electron microscope. Both of these instruments traditionally require a vacuum for operation. It is also likely that either the detection material or the drug material in the coating of the nanoparticle would contain some biological material which would not survive and be viable after exposure to the vacuum. Hence, the result of the measurement, quality assurance, or metrology approach would limit the usefulness of the measured nanoparticle.

This discussion addresses only one of the challenges associated with the use of nanoscale entities for medical treatments. Figure 5.7 represents the multiple aspects and considerations which must be addressed when the use of engineered nanomaterials (ENMs) are proposed for use in or on the human body. Figure 5.7 can also serve as a roadmap defining the intersection between nanoscience concepts, nanotechnology applications, and traditional science disciplines.

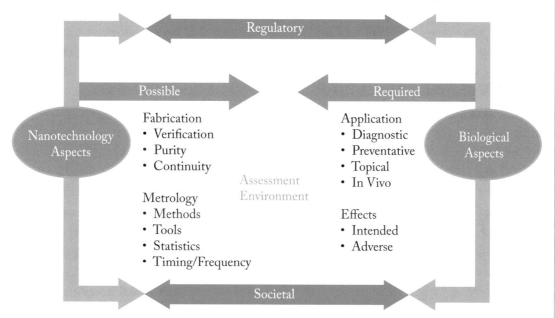

Figure 5.7: Multiple aspects impact and affect the nanoparticle development intended for a medical application.

As shown in the figure, both nanotechnological and biological aspects will impact the manufacture, testing, and application of coated nanoparticles intended for medical applications. Some of the challenges associated with the manufacture or fabrication (see figure) of such entities were discussed previously and include creating the core nanoparticle of the appropriate purity and size. A spherical shape is also desired as opposed to flat sheets or oval-type shapes. The initial interface material needs to attach to both the gold and the detection and drug material. Finally, the attached detection and drug material need to be in the correct percentage of coverage and to remain adhered to the particle for a set period of time.

The measurement approach to ensure viability of the coated nanoparticles is also a variable in the process. The methods of measurement can include visual at different magnifications using optical instruments, chemical testing used to verify presence of the various required elements, and use of nano tools such as scanning electron microscopes. The measurement approaches may also consist of a combination of multiple methods used at different stages of the process. The statistics required as well as the timing and frequency of the measurements will depend upon the particular application. The requirements may result in specific equipment or personnel requirements at the application facility. Each of these restrictions can result in additional cost.

On the right side of Figure 5.7 biological aspects are listed. These factors will have a significant impact on the other factors listed in the figure. If the application is diagnostic, then more than likely the test will be performed in a lab environment using beakers and test tubes. In this situation the restrictions upon the properties of the nanoparticles as well as the measurement methodology will be more lax than in the other applications. The next two applications, preventative and topical, can require varying degrees of nanoparticle integrity and measurement requirements with topical being the less restrictive. Finally, an application where the coated nanoparticle will enter the body under treatment will require the most stringent physical property and measurement requirements. In this situation the method used to introduce the coated nanoparticle into the patient will also impact the other factors and requirements. The coated nanoparticle could be included in pill form, injected or intravenously introduced. Each one of these methods will require differences in the nanoparticle composition and size and the metrology approach, timing, and pass/fail criteria.

In this example an understanding of nanoscale interactions, phenomena, measurement methodologies, and manufacturing approaches is necessary as well as an understanding of physics and chemistry topics and biology. There is a clear overlap, integration, and intersection between nanotechnology and the traditional sciences. Also, there is a need for regulatory, quality assuance, metrology engineers, and social scientists to be aware of nanotechnology influences and requirement within those areas of expertise.

5.5   MULTI-DISCIPLINARY ASPECTS OF NANOTECHNOLOGY

Finally, as exemplified by the above topic, nanotechnology is multi-disciplinary. This not only applies to the integration of nanoscale concepts with the theories and concepts of traditional sciences but also applies to the multi-disciplinary aspect between the traditional sciences. That is, chemists need to know aspects of physics and biology in addition to nanoscale concepts. Similar requirements are placed on physicists and biologists.

This aspect poses a challenge for both businesses and education. To truly understand the world at the nanoscale, aspects of chemical engineering, material science, biology, and other disciplines must be considered. It is clear how the traditional sciences feed into understanding the world at the nano scale. Different disciplines will be required for different market segments or industries where nanoscale understanding can be applied. Finding employees that have fundamental understanding of multiple disciplines is a challenge for businesses. This often results in the need for different businesses to partner either with educational institutions and researchers or with other companies. It also provides a challenge to the companies in the sense that dependent upon their product and their area of expertise they will need to be aware of developments at the nanoscale to provide continuous improvement for their product.

From an educational standpoint, this provides a challenge for professors with areas of expertise that are limited or expertise in one or two areas. The emerging corporate requirement is for employees to understand multiple disciplines and to include multiple disciplines in the understanding of nanoscale phenomena. A review of any of the peer-reviewed technical literature will show multiple authors from not only multiple educational institutions but also corporate employees.

An overview of the technology progression, disciplines involved, and funding sources is shown in Figure 5.8. Across the top of the figure the technology maturity phases are shown as columns labeled 0–7.

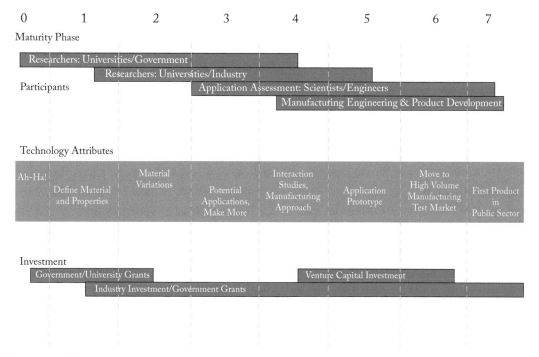

Figure 5.8: Phases as a technology matures showing participants and funding sources.

The middle portion of the figure, shown in green, represents possible attributes for the activity that occurs in the different phases. These phases and attributes are similar to what occurred with the discovery and study of CNTs shown in Figure 4.4. Usually the initial discovery is a conundrum of discovery and trying to understand the basic properties or capabilities of the new material or process. As the discovery process moves forward into phases 2 and 3, more understanding of the discovery occurs, variations may be studied, and the investigation of high-volume manufacturing is initiated.

As the "discovery" moves into the latter phases of the maturity model, applications are proposed, a prototype may be created, test market trials performed, manufacturing perfected and eventually the product is introduced into the public sector. Figure 4.2, which shows the S-curves, also represents the movement from discovery to a money-making product, as shown in the maturity phases of Figure 5.8.

The top portion of Figure 5.8 shows the variety of participants, in general terms, that are involved in the various maturity phases. Initially, in the early phases, scientists and researchers from multiple disciplines, traditional sciences, and engineering segments are involved. This is truly the nanoscience portion of the work. In some cases, educational institution researchers will partner with industry researchers. In this instance, the company "partner" may provide funding for the research, provide personnel to support the work, and often require sole access to the discovery and any products that may result. As the discovery moves on through the phases, personnel with other areas of expertise become involved. This includes application/mechanical engineers, material science engineers, manufacturing experts , and quality control scientists. In the last phase, business development, sales, and marketing personnel become involved. Even in the later phases a certain degree of technical understanding is required in order to understand the nuances and uniqueness of the "new" product. Throughout the latter technical discovery phases, environmental, regulatory, and legal aspects must be considered with the inclusion of personnel in those areas of expertise.

The lower portion of the figure represents funding sources as a "discovery" moves through the phases. Initially, in the U.S. and indeed in most countries, the initial phases are funded by government agencies and research institutions such as the National Science Foundation in the U.S. This funding is usually in the form of grants that are awarded after a competitive process of proposal submission and evaluation. After the very initial discovery, some industry funding may be available again usually with some ownership considerations involved. As the discovery moves further through the phases and into the later phases where a potential application is defined or a viable product has been developed, outside investors such as venture capital companies or angel investors may provide funding. However, these types of investors usually need statements that define the applicable market segment, product specifics, profit potential, and other supporting financial information. As a result, these types of funding sources are usually only available in the later phases of the maturity model. As shown in Figure 5.8, there can be a gap in funding between phases 2 and 4. This is a period of time where government-funded grants usually are reduced in magnitude and may be harder to receive because the work has gone beyond "science" which is a common requirement for government funding. This region absent of funding is often called the "Valley of Death." This is the region of technology maturity where many small or start-up companies can run out of financial resources. Unless anticipated and planned for, this financial lack can cause the end of the enterprise.

It is clear that there is a strong relationship between traditional science and nanotechnology. This relationship will strengthen in the future resulting in challenges for students, educators, regulatory organizations, and industry.

CHAPTER 6

Nanoscience, Nanotechnology, and Engineering

6.1 INTRODUCTION

Engineering has always had its own "stovepipe" in the educational world and is built upon the foundation of the traditional sciences. Engineering typically will deal with the application of the science found in physics, chemistry, etc. Because of this differentiation, engineering must expand the concepts of nanotechnology (no pun intended) to include assembly, testing, quality, safety, and manufacturability. Engineering requires the evaluation of the impact of nanoscale concepts on the realization of a specific product. The chapter will cover electrical engineering (one of the most prevalent engineering disciplines), civil engineering, and mechanical engineering.

6.2 ELECTRICAL ENGINEERING

The devices that drive our lives; the processors and memory devices are based on semiconductor materials and electrical fields. This has been the case for over 60 years. Recent developments, many based on nanoscale understanding, are pioneering multiple changes in electronics. These changes include smaller transistors resulting in more powerful processors and denser memories. These advances, in turn, will result in enhanced artificial intelligence capabilities, better communication, efficient sensors, safer transportation, improved medical diagnostics and treatments, and a myriad of other application related advancements. The potential to use light, photons, rather than electrons, to store or convey information as well as quantum systems is also on the horizon. Knowledge of the nanoscale is also allowing for the investigation and understanding of semiconductor devices, the physics that controls them and the chemistry of the manufacturing process. This understanding will allow the introduction of new materials and less toxic manufacturing chemicals. Electronic device manufacturers crossed the "nano" threshold around 2000 when Intel created a transistor that had 90 nm dimensions. Semiconductor manufacturers have been driven to smaller and smaller devices as the public, industry, and government continue to desire more memory and processing capability as well as higher speed operation.

As electronic devices such as transistors become smaller, the industry has been forced to be on the forefront of nanoscale design and manufacturing. And, as has been historically the case with many nanoscale discoveries, the design approaches and lessons learned by the semiconductor industry are becoming applicable to multiple industries including diagnostics, medical devices, lubricants, building materials, and even paint.

As electronic devices continue to shrink, the manufacturers are reaching the limits of what is possible to create based on the current approach of device manufacturing. This device shrinking is known as Moore's law which correlates transistor size and hence number of transistors per device or within a given area versus time as shown in Figure 6.1.

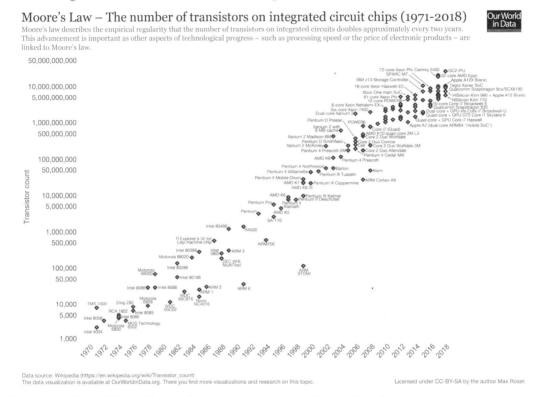

Figure 6.1: Moore's Law: The number of transistors on a chip doubles about every two years.

As the transistor count follows Moore's Law, transistors, by default, are becoming smaller which increases optimal circuit density, performance , and operating frequency. Secondary results are that the power required per chip is also increasing as is the heat dissipated per device. These implications are shown in Figure 6.2. The ramifications can be significant. Providing power to the individual devices and removing the heat are becoming significant challenges as the device geometry is reduced. Device temperature directly influences the reliability and operational lifetime of a circuit, with a negative impact. That is, the warmer the device, the lower reliability and shorter expected lifetime. Often redundancy, either on the chip, where transistors are powered off for a portion of the lifetime, then turned on later or off chip with redundant devices used at the system level. Again, at the system level some of the devices are not operational during a portion of the lifetime but turned on later while the earlier used devices may be turned off, hence keeping the overall system power profile minimized.

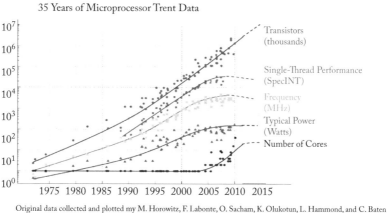

35 Years of Microprocessor Trent Data

Original data collected and plotted my M. Horowitz, F. Labonte, O. Sacham, K. Olukotun, L. Hammond, and C. Baten
Dotted line extrapolations by C. Moore

Figure 6.2: Microprocessor trend data including performance and power based on Moore's Law.

Advances in nanotechnology-based equipment and understanding have had a significant impact in electrical engineering programs and electronic devices. The impact originates in and continues to the fabrication process for electronic devices.

The rule of thumb, previously discussed in Chapter 3, applies to the process of semiconductor fabrication. That is, since the electronic device fabrication approach depends on photolithography—using light to create patterns—the size of the smallest transistors that can be created is approximately equal to one half of the wavelength of the light used to illuminate the pattern. Hence, the semiconductor industry historically has moved to the use of shorter and shorter wavelengths of light to create the devices.

Because the electronics industry is reaching a threshold regarding the size of transistors that can be fabricated, the semiconductor industry is looking at new and novel approaches for the creation of the next generation of "electronic" devices. These new approaches include nanoscale considerations.

6.3   TRANSISTOR OPERATION

The cross section for a MOSFET (nmos type) is shown in Figure 6.3.

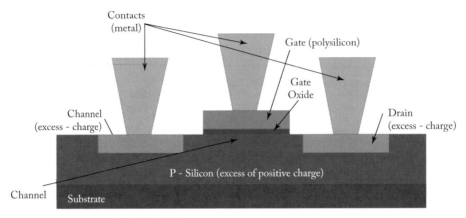

Figure 6.3: Cross section of a MOSFET.

A Metal Oxide Substrate Field Effect Transistor (MOSFET) operates by allowing a charged entity, either an electron or proton, to move from one side of the transistor to the other side causing a current to flow. In the figure, which represents an n- (negative) type device, electrons, which are negatively charged, move from one side of the device (here the left side is designated "source") to the right-hand side of the transistor, designated drain. The region within the substrate underneath the gate is called the channel region and represents the path that charge carriers will move from the source to the drain region. In many cases the distance of the channel region from the sources to the drain, called the transistor width, is used to define the "size" of the transistor. Shorter transistor channels correspond to faster switching of the transistor from on to off and vice versa resulting in increased operating frequency.

The overall operation of the transistor is controlled by the voltages that are applied to the source, drain, gate, and substrate. Voltage is applied to the semiconductor regions by the metal contacts to the source, drain, gate, and substrate regions.

6.4   SEMICONDUCTOR DEVICE FABRICATION

The process used to create an electronic device, called an integrated circuit, for your computer, smartphone or television is a complicated process involving hundreds of steps and a substantial amount of very expensive and unique equipment. The majority of current integrated circuits are structured using a silicon-based process where silicon forms the substrate and the foundation of the electronic device. Other materials may be used as a base, such as sapphire. This base material is often coated with silicon. Gallium Arsenide (GaAs) is another semiconductor that has potential for use in photonic devices.

The birth of an integrated circuit begins with a single crystal of silicon. That initial single crystal is placed on the tip of a stiff probe or wire and inserted into a molten bowl of silicon, as shown in Figure 6.4.

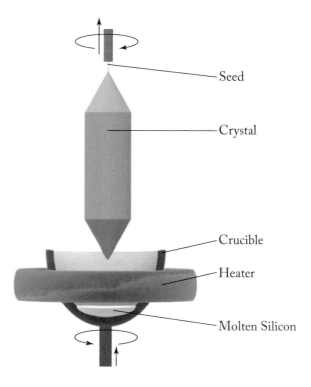

Figure 6.4: A single crystal of silicon being pulled from a molten pool of silicon.

The container for the liquid silicon is most often quartz and the silicon is heated to a temperature of 1425°C (2597°F). The wire with the single crystal of silicon is rotated and slowly pulled out of the molten silicon resulting in the growth of a large single crystal of silicon, called a boule. This process is known as the Czochralski process, discovered by a Polish scientist, Jan Czochralski, by accident. He dipped his pen into molten tin rather than the inkwell which resulted in a single crystal of tin.

The Czochralski process has been used to grow large single crystals of silicon since the 1950's. Initially, the crystals grown were 18–22 cm long and only several centimeters in diameter. Through experimentation and multiple process improvements, silicon boules now are 1 m or more in length and 30 cm in diameter, weighing hundreds of pounds.

The silicon boule is then sliced into thin wafers. The wafers form the foundation and substrate for todays integrated circuits. The fabrication process begins after the silicon wafer slices are smoothed to a uniform thickness and polished. As can be seen in Figure 6.3, there are multiple materials and layers that comprise an electrical device.

The fundamental process steps are shown in Figure 6.5. The first step is to grow an oxide layer on top of the silicon substrate. This oxide layer will eventually become a layer which can act like a capacitor between the active regions of the device: the source, channel, and drain regions and the controlling regions of voltage application, such as the gate. This growth of the oxide layer is shown as Step 1 in the figure.

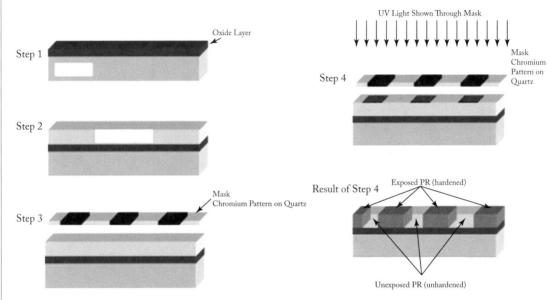

Figure 6.5: Initial fundamental process steps for electronic device fabrication.

The next series of multiple steps involves "growing" or placing specific entities such as dopant atoms, ions with excess electrons or protons, polymers, or metals in specific patterns on top of the oxide and substrate. To obtain the desired specific patterns a photolithography process is used. That is, "light" (photo) is used along with specifically designed masks (lithography) to create the patterns.

After the oxide layer has been grown the entire wafer is coated with a photoresist material, as shown in Step 2. Depending upon the material type, the photoresist will either become very hard or become very soft upon exposure to light. The mask, shown in Step 3, with the specific transistor or circuit design is usually a quartz material with the pattern placed on the glass using chromium.

The creation of the chromium pattern is based on the intended function of the electrical device which, in turn, defines the transistor configuration as well as transistor size and interconnect pattern. The pattern is created based on the electrical engineering circuit design. Figure 6.6 shows the complex equipment used in the facility to create the chromium pattern on the quartz plate.

The equipment shown in Figure 6.6, as well as all of the equipment used in the hundreds of steps required to create an integrated circuit and the personnel are contained in a "clean room."

A clean room is a specially sealed and designed facility with the sole purpose of keeping all dust, dirt, contaminants, or foreign material out of and away from the wafers and circuitry being fabricated. This is because any foreign entity which may become integrated in the device will render that circuit inoperable or unusable. The percentage of operational devices on a given wafer is known as the "yield" for the wafer. Any contaminants that infiltrate the electrical circuit, at any step, will impact device operation and hence reduce the yield, which in turn will reduce profitability of that wafer and the electronic circuits on it.

Figure 6.6: A view into a semiconductor fabrication clean room.

Personnel that work in the clean room are also required to reduce the amount of potential contaminants such as dust, lint, and skin particles that enter the clean room and therefore are required to wear specifically woven and designed suits, which include foot and head coverings as well as gloves and sometime face masks. The clean rooms are equipped with special filters and the air flow is from the ceiling down and through the floor, again with the intent to remove and control any particulates in the clean room facility.

Returning to Figure 6.5, in Step 3 the mask is aligned over the wafer that has been coated with photoresist. In the next step, light of a specific wavelength is used to illuminate the mask and shine onto the photoresist on top of the wafer. The circuit design on the mask, the wavelength of light used, and the spacing between the light source, mask, and photoresist are all dependent upon the specific circuit being fabricated, its geometries, and material composition. These fabrication parameters will change as the fabrication process continues because different layers of the electrical structure will have different geometries.

The lower right-hand portion of Figure 6.5 shows the result of Step 4, where light of a specific wavelength is shown through the mask. In this example, the photoresist which was exposed to the light became hardened and the portion of the photoresist which not exposed to the light, or in the shadows of the mask, is unhardened. In the next step, not shown, the unexposed, softer photoresist is washed away by a specifically formulated chemical bath. The results of the chemical bath, shown in Figure 6.7 as Step 5, result in regions of hardened photoresist on top of the substrate and oxide layer. These regions of hardened photoresist will serve as barriers, when, as shown in Step 6, specific ions are implanted into the oxide layer. It is the implanting of these ions with excess or absence of electrons that result in the desired electrical characteristics of regions of the circuitry or transistor such as the source and drain regions.

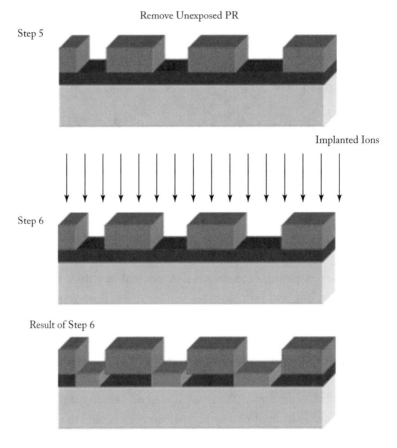

Figure 6.7: Ion implantation steps.

Figure 6.8 shows some of the equipment used for the ion implantation stage. There are different pieces of equipment and different process steps required for photoresist deposition and removal, mask design and creation, ion implantation, metal deposition (for the electrical contacts), and various etching steps.

Figure 6.8: A subset of the equipment used for ion implantation.

These steps are repeated multiple times to create the very complex layered structure of a semiconductor-based device. These transistor structures which result in electronic circuits often end up with tens of layers. The layers are needed not only to provide power/voltage to the transistors to control operations but also provide the interconnecting architecture between all the transistors and functional regions of the device. A schematic of the devices and interconnect is shown in Figure 6.9.

The discipline of electrical engineering goes beyond the task of creating an electronic device that can be used in a multitude of applications. Although nanotechnology has had the most profound impact on the actual fabrication of electronics this impact extends beyond the fabrication process. At the conclusion of the many fabrication steps, the product is a silicon wafer with hundreds if not thousands of electronic circuits patterned on the surface. The wafer is then sawed, usually with a diamond saw, to separate the individual devices. At this point the individual devices are vulnerable to dust, particulates, and any type of contact that could damage the circuitry which is exposed on the top surface of the device.

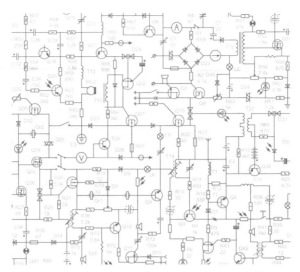

Figure 6.9: An electronic schematic representing the various electronic devices and complex interconnect required.

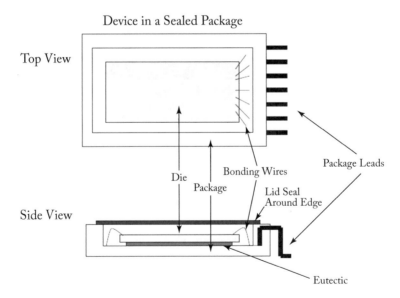

Figure 6.10: Electronic device in a sealed package.

To make the circuit usable and able to be handled, the device must be placed in a package. The package not only protects the device from the environment and any contaminants or physical threats but also serves to connect the electronic circuitry to other devices and the "outside world." The task of designing the device package, establishing the configuration of the connections to the outside world, and placement of the packaged device on a circuit board where it will connect to other devices is also the role of an electrical engineer. In each of these tasks nanotechnology, especially in material selection, has begun to impact package and circuit board design. At this point in the process engineers with mechanical or material expertise may also be involved.

Figure 6.10 shows a top view and side view of an electronic device in a package. The electronic device is adhered to the package using a eutectic that could be either a metallic-type solder or a polymer epoxy. This path from the device through the eutectic and into the package is most often used for the removal of heat from the device. Eutectics which include nanoparticles of specific high thermal conductivity can be used to support the removal of heat from devices as the transistors become smaller and more heat is generated per device.

The periphery of each electronic device includes bonding pads. These bonding pads are then connected to landing bars in the package. These landing bars are then connected to the pins on the outside of the package, as shown in Figure 6.10. The materials used in the package, bonding wires and pins to the outside world are all being impacted by material-based nanotechnology research.

Finally, after the device is securely placed and sealed in a package, that package is mounted on a printed circuit board (PCB). At this point in time individual devices such as a Central Processing Unit (CPU), are connected to other devices such as input/output devices or memory devices via a multi-layered board. The layers within the board have traditionally been made of copper with via connectors between the layers within the board; see Figure 6.11. Again, the material and structure of the PCB are designed for mechanical stability, heat transfer capability, and electrical functionality.

This process involves multiple engineering disciplines including electrical engineering, mechanical engineering, manufacturing engineering, and reliability engineering, to name a few.

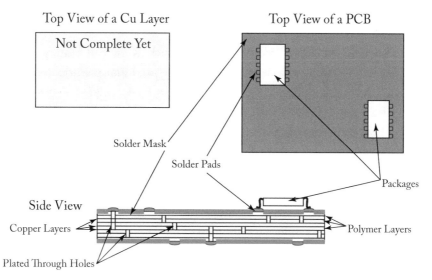

Figure 6.11: Multi-layered printed circuit board (PCB).

6.5   MECHANICAL ENGINEERING AND MATERIAL SCIENCE

These two disciplines are often closely related especially as they apply to nanoscale phenomena. Mechanical engineering is a broad discipline that includes mathematics, physics, material science, and other disciplines related to the design, analysis, manufacturing, and maintenance of mechanical systems.

Figure 6.12: Ball bearings: resulting from mechanical engineering tasks.

Mechanical engineers design systems that move such as the ball bearing wheels shown in Figure 6.12 to engines, industrial manufacturing machinery, and transportation vehicles from lawn mowers to ocean going oil freighters. Material science can be considered a subset of mechanical engineering, or vice versa, and deals with the composition of materials at all size scales and the physical and electrical properties of those materials.

Mechanical engineering, because of the diversity and variety of tasks involved in the discipline, can be and is significantly impacted by developments at the nanoscale. One significant area of impact is the use of nanoscale coatings. Because many of the products designed by mechanical engineers include moving parts, nanoscale coating can be used to reduce friction and/or remove heat. Both friction and heat can reduce the lifetime of a unit, hence reducing those factors by the use of a coating can provide a financial benefit.

Materials based on nanotechnology, which may use nanoscale materials, will have different physical properties than the large macroscale counterparts. For example, nanomaterials may respond to water environments differently than material without nanoscale designed structures. The nanoscale coatings can result in providing either hydrophilic or hydrophobic surfaces. Nanomaterials may also provide more flexibility or ductility than larger materials or potentially interact differently with various chemicals because of the difference in surface energy. Nanocomposites, which can include nanoscale particles of clay, CNTs, or nanofibers, can significantly impact, in a positive manner, the thermal properties, weight, electrical properties, mechanical strength, and reliability of the resulting application or product.

Hence, nanotechnology through mechanical engineering and material science disciplines has applications in the aerospace and automobile industries, anywhere coatings are used or required, in the energy sector improving efficiency and in building materials adding fire retardation and corrosion protection.

6.6   CIVIL ENGINEERING

The profession of civil engineering deals with the design and construction of natural and man-made structures. This includes structures such as roads, dams, and sewage systems, as well as canals, canyons, and natural bridges. Civil engineering involves the analysis and understanding of structures and material science coupled with knowledge, analysis, and understanding of soils, geology, hydrology, and phenomena such as hurricanes and earthquakes and how they impact structures, both natural and man-made. An example of a civil engineering designed and built structure is the dam shown in Figure 6.13.

Figure 6.13: A dam: representative of civil engineering tasks.

In designing this structure, civil engineers had to consider the physical requirements for the dam such as water pressure on the structure, load considerations for the top of the structure, as well as consideration for the internal equipment used for energy generation and how that equipment may impact the physical capability of the dam.

Civil engineering also involves assessment of natural structures such as the natural bridge shown in Figure 6.14. Working with geologists and material scientists, civil engineers will assess the integrity of the natural structures with regard to safety and longevity in various environments. Hence, civil engineering includes both natural and man-made structures.

Figure 6.14: A natural bridge.

Civil engineers, like engineers from other disciplines, have learned lessons from nature and are applying them to man-made structures. An excellent example of this relationship is the abalone. The abalone is a large bi-valve mollusk in the same biological family as mussels, oysters, and clams. Abalone are one of the larger members of this phylum of Mollusca and can measure between 12.5 and 18 cm, weighing approximately 350 g.

Sea mollusks and abalone are some of the favorite meals of sea otters and they often will float on their backs in beds of kelp using rocks to break open the mollusks; see Figure 6.15.

A chemical assay of the composition of the shells results in a composition predominantly calcium carbonate, $CaCO_3$, which is chalk. Most everyone is familiar with chalk and is also aware that it is a relatively soft material. For many decades, the known composition of the "soft" material of the abalone and other mollusk shells did not match the required energy and force that the otters had to exert to break them open.

Figure 6.15: A floating sea otter.

Over the last 20, years scientists from varied disciplines such as physics, chemistry, and molecular biology have been studying the abalone and other shells using nanotechnology tools such as AFMs and SEMs. The initial discovery, in 1999, was that the layers of the inner portion of the shell which are similar to the bricks in a wall were separated by a biologically based polymer material that acted as a flexible mortar. This flexibility allowed the shell layers to withstand applied force without breaking or shattering. The discovery of this flexible layer was the first of seven architectural features of the abalone shell, understood through research spanning several years, that contribute to the unusually high strength of this animal's shell.

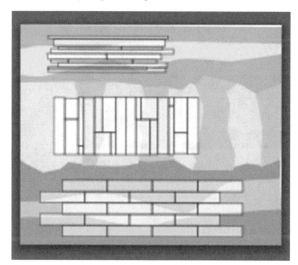

Figure 6.16 is an artist's image of a cross section of an abalone shell in which the layers of material can be observed. The lower portion of the drawing represents the brick-like structure discussed above. In the middle of the image is a set of structures that are columnar in shape. On the top of the cross section is a layer of horizontal plates, which also contribute to the strength of the shell.

It is these various layers with different geometric structures coupled with the flexible polymer separation material that contributes to the strength of the mollusk shells.

Figure 6.16: Artist's drawing of the cross section of an abalone shell.

Based on these discoveries and understanding the material and feature contributions of the shell that result in not only its strength but also contribute to the response of the shell to various stresses, civil engineers are integrating some of the discovered architectural features into the design of buildings which are being built in earthquake prone regions. It is intended that by implementing these features into large structures that susceptibility to earthquakes will be significantly reduced. This architecture of layers of horizontal and vertical structures is also being applied to create lighter-weight bullet-proof vests.

6.7   CONCRETE, CIVIL ENGINEERING, AND NANOTECHNOLOGY

Concrete is one of the most, if not the most prevalent, of building materials worldwide. Concrete is composed of an aggregate material, water, and a cement material, which itself can be a liquid. Through the mixing of these and other materials the concrete is created and poured into the desired location or form. Figure 6.17 is an example of the use of concrete over multiple decades in the U.S.

Figure 6.17: Concrete is the most prevalent consstruction material used. The Tunkhannock Viaduct in Pennsylvania is an example.

The wide use of concrete as a building material has substantially contributed to air pollution, especially in China, where the use of concrete is significant. It is not the concrete itself which contributes to the pollution but the creation and manufacture of the required cement. This catalyzing or binding agent, cement, is necessary in the formulation of concrete but is having a devastating impact on the environment.

Researchers studying the impact of cement manufacturing, as well as the role it plays in concrete use, stability, and properties, have begun to look at nanomaterials as alternatives to the cement material. One of the alternatives is CNTs. Since the CNTs for this application do not require any electrical characteristics and the primary purpose is structural, it is not necessary to create CNTs of an exact or identical nature. Therefore, CNTs of this type can be manufactured in relatively high volumes making them cost effective. It has been found that a very small percentage of CNTs, when mixed in with the aggregate and a fluid such as water and a smaller amount of the cement, are used to create the concrete that a similar concrete strength can be achieved while using up to 50% less concrete. That is, a concrete function that would normally require a concrete thickness of 10 cm could be achieved with a thickness of 5 cm.

In addition to CNTs, other engineered nanoparticles (ENPs) are being investigated as additions to concrete that will take advantage of the unique properties of the ENPs. For example, titanium dioxide nanoparticles are hydrophobic, repelling water. When added either into the concrete or used as a coating, water is repelled from the concrete. Water sinking into any cracks or deformations of the concrete can undermine the structural integrity of the material. Therefore, keeping the water away from the concrete, or removing it quickly, will increase the longevity of the concrete. Titanium dioxide nanoparticles can also be used as a coating for glass, resulting in self-cleaning windows.

Finally, using a capsule, nano-encapsulated epoxies, or resins can be integrated into the concrete as it is mixed. The capsules remain intact in the concrete until a crack, usually a small one occurs in the concrete and breaks the capsule. The internal resin is released and seals the crack.

A word about engineering disciplines, nanotechnology, and higher education credit requirements. As can be determined by the examples in this chapter as well as other sources of information, nanotechnology and understanding the world, systems and materials at the molecular and atomic level will have significant impacts for all technical and engineering disciplines. However, it is acknowledged that these technology-based disciplines have a substantial amount of content that students are already required to learn. In many states, a credit "upper limit" is being placed on engineering and science departments. Typically, this upper limit is 120 credits for a quarterly based educational system and a four-year program. The rationale for this upper limit is to allow students to obtain a degree in a four-year time period. Educators are being challenged by this 120-credit limit to be able to include basic curriculum and topics directly related to the technology of the program. The introduction of nanoscale content and its application and impact poses an even greater challenge. Educators, students, administrators, and researchers are working together, as well as government agencies such as the National Science Foundation and the Department of Education, to address this issue. One approach for students is to apply for positions in summer nanotechnology-based research experiences, often at universities and supported by the National Science Foundation, to supplement the traditional course content taught during the four-semester school year.

CHAPTER 7

Emerging Technologies

7.1   INTRODUCTION

Many of the technologies that fall under the umbrella of emerging technologies are a result of an emerging understanding of the world at the molecular and atomic level. The "umbrella" of emerging technologies includes, for example: photonics, composite materials, biotechnology, additive manufacturing, and sustainability. These technologies are a result of and are influenced by the concepts and discoveries at the nanoscale. The converse is also true in that understanding the nanoscale is dependent on understanding photonics, biotechnology and so on. Often, these emerging technology topics are used in combination and applied to markets such as construction, agriculture and appliances. This chapter will look at photonics and biotechnology and provide a broad overview of this integrated relationship.

7.2   NANOTECHNOLOGY AND AGRICULTURE

Nanoscience, in combination with some of the other emerging technologies, is influencing sensor systems in multiple markets. In particular, there is recent interest is the agricultural market to use nanotechnology for sensor applications. The challenge of feeding the ever-increasing world population with nutritious food continues to grow. The issues include finding a sufficient land mass required to grow the volume of crops required, providing the needed nutrients and water for the plants, as well as controlling insect and biological infestations.

Historically, the approach used to increase both the yield and the nutritional value of food crops, for animals and humans, has been selective or hybrid breeding. Addressing specific plant needs and environmental and pathogen-related malignancies has traditionally relied on the observations and awareness of the land manager. Obvious solutions included irrigation, soil fertilization, and chemical insecticide or fungicide distribution. The problem with many of these solutions, although effective, was that there was a substantial amount of wasted material, whether it was water or some distributed chemical. This was often the case because, for example, plants that did not need water, also received water and areas of the crop that may not have an insect infestation still received the insecticide.

In the early 2000's, nanotechnology-based sensors were being used by vineyard owners in Australia to measure the moisture level of the soil. The sensors were fitted with a radio frequency (RF) transmitter and the vineyard manager could move through the vineyard and remotely access the soil moisture content being measured by the sensor. In this manner, only portions of the vineyard that needed water were indeed watered. In the U.S., satellite imagery has been used to determine insect- or fungal-infected regions of large fields in order to focus treatment only on the areas that needed to be treated. Although these methods were effective, they were only effective to a certain degree and tended to be expensive.

More recently, with the understanding of various plant genomes, genetically modified (GMO) versions of plants such as wheat and corn have been used to improve the properties of these crops. The public fear and outcry against GMO plants or foods has created a roadblock in terms of implementing this approach to its fullest potential.

The ideal situation would be if the plants could be read directly with regard to the overall health and status of the plant. Very recently, researchers have been creating nano-based sensors that can be placed directly on the plant. These sensors are being developed to measure not only moisture/water content within the plant but also levels of other chemicals necessary for the successful life of the plant. The chemicals to be measured, of course, depend on the specific plant type. Other sensors which can be placed on the plant can be used to measure the presence of undesired entities such as insects, fungus, and bacteria.

A predominant approach being used to "read" these sensors is though optics or photonics. The sensors are designed so that under "normal" circumstances they are stable and remain neutral in color or provide a null reading in the detector. In the presence of the "to-be-detected" entity, the sensors may fluoresce and change color, producing a directly observable result or emit a signal that can be detected by a remote or removed device.

7.3   PHOTONICS

Photonics is the science that deals with light. According to the National Science Foundation funded center, OP-TEC, The National Center for Optics and Photonics Education, "Photonics is the technology of generating and harnessing light and other forms of radient energy whose quantum unit is the photon."

Photonics deals with lenses, lasers, transparent materials, and fibers. Lasers, first developed in 1960, are perhaps the first representation of the photonics field. Most often photonics deals with visible light or light that is very close to visible light. Visible light constitutes a very small portion of the electromagnetic spectrum, as shown in Figure 7.1 and encompasses wavelengths from 400 nm (blue) to 700 nm (red).

Our vocabulary changes for shorter wavelengths to discussion of radiation and X-rays and gamma rays and it also changes as the spectrum encompasses longer wavelengths to sonar and radio frequencies.

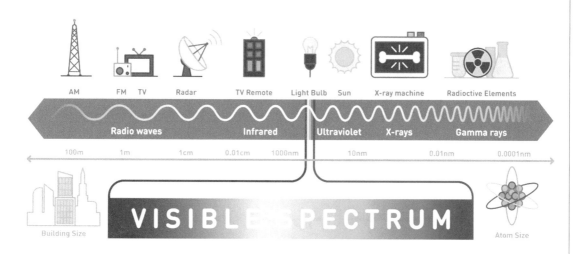

Figure 7.1: The electromagnetic spectrum (VectorMine/depositphotos.com).

As is well known, light, or any portion of the electromagnetic spectrum, can be viewed as a particle or a wave. This is known as the wave-particle duality of light. The name for the particle version of light is photon, hence the name of this technology; photonics. A common application of photonics is fiber optical communication. Where, rather than conveying information in the form of electrons moving through metal wires and generating currents, voltages, and other electrical phenomena, information is contained in the form of light packets and transmitted through "cables" formed out of non-electrically conductive transparent medium. Currently, at both the initial and final end of the fiber optic cable, information, starting in the form of electrons, must be converted into photons and then converted back again to electrons for further interpretation and transmission.

It has long been the desire to combine the circuitry that defines an electron-based technology with equipment that supports a photon-based technology. This has proven to be a challenge because the predominate technology which supports electronics is a silicon-based technology and a gallium arsenide technology supports a photon based technology. Although similar, the thermal characteristics of the two materials differ sufficiently that when layered together the mismatch between contraction and expansion of the materials over a small temperature range will cause the two materials to become separated. Once separated, the "communication" between the electrons and the photons is no longer possible.

Nanotechnology and light or photonics have always been closely related and even more so in the present day. An interesting example is the blue morpho butterfly shown in Figure 7.2. This butterfly has beautiful, iridescent blue wings and, as with most objects, it was thought that the color in the wings was due to a chemical acting as a dye in the wings. However, recently using the tools of nanotechnology such as the AFM and SEM it was determined that the wings of the butterfly

Iridescent colors of *Morpho* butterflies are caused by the specific nanostructures on their wings (SEM image at the bottom).[6]

contain a tree-like structure. The "branches" in this tree-like structure are spaced in such a way that they only reflect light in the blue wavelength.

That is, there is no dye or chemical aspect to the color of the wings but it is dependent upon the physical structure in the wing. It is now being discovered that peacock feathers and other natural objects appear a certain color due to the nanoscale physical structure and how that structure interacts with light more than a dependence upon chemistry.

Figure 7.2: The Blue Morpho butterfly and an SEM image of the structure in the wings.

7.4   BIOTECHNOLOGY

Biotechnology is defined as the exploitation of biological processes for industrial and other purposes, especially the genetic manipulation of microorganisms for the production of antibiotics, hormones, etc.

Biotechnology, as the name implies, is a combination of the science of biology and the application or developing the technology which results from that science. Often phrases like molecular engineering, biomedical engineering, and molecular biology are used to describe more specific areas under the broad umbrella of "biotechnology." Biotechnology refers to the application of biological processes for the development of pharmaceuticals, antibiotics, and so on. Biotechnology as a category also includes the work that involves understanding genes in humans, plants, and animals and their roles in health and diseases.

Initially, biotechnology was a subject material that was included in traditional classes such as biology, ecology, and various pre-med courses such as detailed studies of the various human systems, for example, the cardiovascular and immune systems. As the tools of nanotechnology were developed and researchers were able to investigate biological systems at greater level of detail, the subject matter covered under the biotechnology umbrella became vast enough that it grew into a standalone discipline. Advances in nanotechnology have and continue to support the advancement of biotechnology.

For example, one aspect of biotechnology involves the analysis of blood. When blood is assessed various quantities are measured such as red and white blood cell count, and iron, sugar and salt concentrations. Physicians may also request additional tests if other diseases or cancers may be suspected. When a test is run to look for a specific cancer, the test may not be identifying actual cancer cells but rather tumor markers or proteins that are expressed or created by the cancer cells or tumor. Each type of cancer appears to have a representative protein that is specific to that cancer. These proteins are called target proteins. Hence, when blood is drawn it and then tested for a specific target protein representative of the suspected cancer type. Historical, traditional tests had several drawbacks. First, they could be inconclusive and take a lengthy period of time to evaluate. Second, usually a relatively large quantity of the protein under investigation had to be present, which meant that the cancer had progressed to a potentially advanced stage.

Using tools of nanotechnology, researchers were able to better quantify and describe the target proteins from a chemistry perspective. Researchers also were able to more clearly identify the characteristics and differences between healthy and unhealthy cells and differentiate between cancer created proteins and healthy, normal proteins. Understanding of these factors was a result of investigation and study at the nanoscale. A new library of tests has been developed using nanotechnology that allows for the detection and measurement of the target proteins in much smaller quantities. This leads to earlier detection of disease, simpler and less expensive treatment, and a better prognosis.

The discussion in this brief chapter represents the proverbial "tip of the iceberg" for nanotechnology research, understanding, and application.

Author Biography

Deb Newberry served as the Director/Instructor of the Nanoscience Technician program at Dakota County Technical College in Rosemount MN from 2004–2018. She created the 72-credit nanoscience technician program in 2003 and began the program with National Science Foundation funding. Deb also served as the Director and Principle Investigator of the Center for Nanotechnology Education, Nano-Link for over 10 years. Nano-Link has been funded by over $12M from the National Science Foundation. Nano-Link educational content, developed by Deb, has been used by over 900 educators and has reached over 100,000 students. She is a nanotechnology book co-author, has written over 12 book chapters and presented more than 250 presentations and tens of educator workshops.

Deb Newberry is currently serving as CEO of Newberry Technology Associates (NTA), a company that provides expertise in organizational structure and efficiency, emerging technologies, strategic planning, technology evolution, and product development. The NTA team performs business and competitive analysis for multiple technologies including nanotechnology, electronics, photonics, material science, additive manufacturing, and biotechnology. By creating an Idea Realization Team, NTA helps to determine how emerging technologies may impact products, companies, and market segments.

Prior to her career in education, Deb worked in the corporate world for 24 years performing thermal and radiation testing and analysis on satellite systems and then serving as Executive Director managing over $450M worth of satellite programs. Ms. Newberry led a national committee with a focus on determining the impact of nanotechnology on satellite electronics and has served on multiple advisory boards as well as state and national commissions. She is a member of professional organizations such as the IEEE, ACS, MRS, RCS, and ASEE and serves on multiple conference planning committees.

Printed in the United States
by Baker & Taylor Publisher Services